I0606188

AUTOBIOGRAPHY OF A SKYSCRAPER

PRAISE FOR FRANCIS J. GREENBURGER

"Francis is a gifted connoisseur of many things. His eye for real estate and his refined taste in art and architecture is truly remarkable. However, it is his ability to identify and inspire gifted people, an ability that stems from his curiosity, admiration, and love of human beings, that is his superpower."

—PATTY MCHUGH
Chairman, McHugh Construction

"The nine-year odyssey of 1000M is a testament to our collective will to make this building the best it can be. Francis continually challenged us to do better. Likewise, as architects, we continually challenged him to push the residential tower to new limits... Through this collaboration and tension, we together elevated the architecture of 1000M to the level of art."

—PHILIP CASTILLO
Executive Vice President, JAHN

"The last 10 years have been the most remarkable business experience of my career. The project was extremely difficult to bring to fruition. In the face of massive personal financial risk, countless complications, business plan changes, Francis remained stoic. I will always be in awe of his vision, strength, and ability to adapt. Without these qualities, there was no possibility of success for 1000M."

—ROB SINGER
Director of Development, Time Equities

"1000M will forever be Francis Greenburger's enduring legacy in the city of Chicago because of its sheer prominence and skyline-altering nature. But under the surface, 1000M is the embodiment of one man's vision and determination to deliver a final product that many deemed impossible for one reason or another. His ability to deliver on time and within budget despite the many headwinds, including a global pandemic, will be the true legacy of 1000M."

—CHRIS PARILLO
Managing Director, Deutsche Bank

"Building new construction, while always a challenge, was exceptionally difficult given the pandemic and seismic changes that we all experienced in the way that we live and work. Francis was able to gracefully lead with a steady hand, creative thinking, and constantly encouragement."

—JORDAN KARLIK
Partner, JK Equities

"In this phenomenal book, Francis Greenburger gifts us with paired capstones of a life lived generously and triumphantly. First, one wrought in glass-and-steel, and second, one crafted in paper- and-ink."

—ROD COLBURN

"What is so notable about Francis as an art collector and advocate is how strongly he integrates his passion for art into his development practice, putting it at the forefront of every project and sharing what he loves with tenants and visitors to these spaces.

His commitment to the arts is remarkable, so much so that he has created a position for an in house curator to work directly with him in realizing a rich art program for each space, large and small."

—TESSA FERREYROS
Curator, Time Equities

"In my collaboration, and partnership, with Francis on 1000M, I found him to understand the magnitude and lasting impact that this building would have on Chicago's skyline. From a design standpoint his vision, aesthetic and attention to detail was extraordinary."

—KARA MANN
Founder and Creative Director, Kara Mann Design

AUTOBIOGRAPHY OF A SKYSCRAPER

The Story of Chicago's 1000M
and Those Who Built It

FRANCIS J. GREENBURGER
with REBECCA PALEY

O/R
OR Books
New York • London

Published by OR Books, New York and London

Visit our website at www.orbooks.com

All rights information: rights@orbooks.com

First printing 2025

The manufacturer's authorised representative in the EU for product safety is Authorised Rep Compliance Ltd, 71 Lower Baggot Street, Dublin D02 P593 Ireland (www.arccompliance.com).

Cover design by Josef Reyes. Typeset by Poco Meloso.

Printed by Sheridan, USA, and CPI, UK.

hardcover ISBN 978-1-68219-516-1 • ebook ISBN 978-1-68219-517-8

Photos edited by Donna Cohen.

Photo Credits:
JAHN - Cover, pp. 1, 6, 11, 13, 15, 23, 25, 26, 27, 30, 98, 99, 106, 160
Tom Rossiter - Back cover, pp. 35, 304, 311
Michael McWeeney- Author headshot
Adam Heneghan - p. 32
KARA MANN - pp. 34, 36, 250
Matt Mansueto - pp. 37, 55, 58, 103, 145, 150, 153, 155, 158, 160, 162, 163, 165, 167, 169, 171, 172, 173, 174, 179, 181, 190, 192, 195, 197, 199, 203, 206, 209, 211, 213, 214, 217, 221, 232, 233, 234, 236, 237, 238, 243, 245. 247, 249, 257, 261, 263, 270, 272, 290, 296, 300, 303
Otherwise Inc. - p. 46
Millerhare - pp. 61, 62, 64, 65, 118
Nam Y Huh/AP Photo - p. 83
DBOX - p.94
Dr. Albrecht Bangert - p. 114
McHugh Construction - pp. 134, 148, 264, 265, 267, 309
Modular Closets - p. 187
Full Bars Media - pp. 228, 229, 251, 252, 254, 255
Pictury - pp. 275, 277, 279, 282, 286
Courtesy of Steve Poliseno - p. 307
Envida - p. 313

For all my colleagues, partners, and stakeholders
who made this dream come true.

CONTENTS

A NOTE FROM THE AUTHOR

The role of a developer is similar to that of an orchestra leader.

As a for-profit apartment building developer, our job is to design and construct buildings that are economically viable. But it's more than that. To be successful, these projects must also be welcomed by local governments and surrounding communities. Our guiding principle is that a development project must work for all stakeholders—developer, government, and community.

The developer must identify a community need, find a site for that need, and convince the community that they have the right vision to fulfill that need, and also the right location within the community. Then, once government approves the plan, the hunt for resources begins. The developer must convince those who control money (banks and investors) to take a risk on the vision and to have the confidence that they have the ability and the experience to realize the vision "on time and on budget."

The developer doesn't work alone. To complete all this "takes a village"—hundreds of people working in perfect coordination and lockstep. This "village" includes professional land planners, architects, engineers, lawyers, land-use consultants, estimators, construction managers, interior designers, tradespeople, material suppliers, site supers, and safety managers. It takes marketing experts, leasing or sales agents, property managers, and ultimately residents who sign up to live or work in the building.

Each step must be completed accurately and on a schedule in accordance with the critical path. If one person or vendor delays, it will have a snowball effect on the others waiting to begin their work.

And, of course, the developer must overcome an endless list of unexpected problems that will affect the progress of the project: a change in government; supply chain delays; design mistakes; contractor mistakes; weather-related delays; water leaks; a pandemic; and on and on.

In this book, we have tried to tell the story of this process, to take you inside this village and this progression, by bringing you many voices of the community that give rise to a major skyscraper.

You will hear directly from members of the village: about their background and experience; their joys, frustrations, and personal commitment to the project; and certain "tricks of the trade" they've generously shared.

For all our readers—whether you're a general-interest reader, developer, architect, engineer, urban planner, civil servant, tradesperson, designer, or simply interested in the growing skyline around us—we hope you will gain a better appreciation for the stories behind the jungle of buildings that many of us wonder about as we live in, work in, or walk by them.

FRANCIS GREENBURGER

SLOPE
1000M

Chapter 1

1000 SOUTH MICHIGAN AVENUE, 2014

A Downtown Chicago Parking Lot Goes On Sale

FRANCIS GREENBURGER

Founder, Chairman, and CEO of Time Equities, Inc.

Jerry Karlik got off the plane from New York in winter 2014 and was instantly met with a blast of Chicago's bitterly cold and snowy weather. The trip had been bumpy with the pilot complaining of icy conditions on the ground. But it was nothing compared to the rough ride the vacant South Michigan Avenue property across from Grant Park had been through in the last six years—and what the real estate developer who had just flown in to bid on it was about to experience.

"Dad, it's a very interesting piece of property," Jerry's son Jordan, the other half of JK Equities, had told his father months earlier.

Interesting was one word for 1000 South Michigan Avenue in the South Loop's Historic Michigan Boulevard District near the park and Lake Michigan. A Chicago-based developer had purchased the site in 2005 for $48.3 million with a plan for a 40-story, 346-unit condo tower, but they didn't even break ground before the market took a dive. The bank that provided the financing foreclosed on the property in January 2009 and took it over though a 2010 auction for just $11.3 million, less than half what the developer owed on the site.

Jerry started out in the early eighties as a real estate tax shelter syndicator, buying properties and syndicating them for the tax benefits. His New York–based company didn't just concentrate its investments in the Northeast but spanned out to the South and the Midwest. In 1997, he developed his first property in Chicago—mostly out of frustration. Dearborn Tower began as a warehouse in the early 1900s that was located in the old industrial section of the South Loop, a run-down neighborhood not unlike the industrial sections of many American cities. Although a residential development was an anomaly for this area, Jerry had a vision to convert the warehouse into a 325-unit condominium—except he couldn't get anyone to do it. Finally the broker working for him said, "Why don't you just build it yourself?" And so he did. Dearborn Tower sold out in less than two years, and a developer was born. While trading real estate can be extremely rewarding and lucrative, creating buildings that will last longer than anyone's lifetime is the stuff of legacy. It is also a very different business.

Jerry, aided by his son Jordan, began to seek out more development opportunities in Chicago. After a stint on Wall Street, Jordan had planned to attend law school. With a little time off before he started, however, he agreed to help his dad in Chicago. Jordan found the work thrilling. Law school fell to the wayside as the Karliks dug into Chicago real estate, which was on a rocket ride up until 2008. "From 2009 to 2011, we were professional firefighters trying to put out problems, some of which we created," Jordan said. "When we weren't putting out fires, we were still trying to find acquisitions to capitalize on the other side of the distress."

The whole period was a great education for Jordan, who made Chicago his home and grew to know a lot of the city's real estate players. It was through one of these connections that he learned about

1006 South Michigan Avenue, an eight-story loft office building. The building's price of $11 million was on the expensive side for the area at the time, but the air rights were worth around $3 million. At $8 million, the building was a really good buy. Even more crucial—and lucrative—the Karliks discovered that when someone came around who wanted to develop 1000M, they would have to purchase from the owner of 1006 not only the air rights but, more importantly, the egress rights for parking. In Chicago, you can't build a condo without parking, and 1000M only had frontage onto South Michigan Avenue. The parcel with the office building extended to South Wabash Avenue, the street parallel to the west where cars *could* drive into a condo tower, overlooking the treetops and beyond that the endless sparkling blue waters of Lake Michigan.

That's when Jerry came to me looking for a partner.

About 40 percent of the deals at my company, Time Equities, Inc., are joint ventures. Partners will often bring us local knowledge of the real estate market or access to a transaction. Often we complement them with not only money and bank credibility but also a full spectrum of skills and experience that come from more than fifty years in the real estate business.

Founded in 1966, Time Equities started out as not much more than me hustling to make small profits off of properties in New York City's rental market that went from difficult to impossible during the city's financial crises of the 1970s. Pivoting to the transformation of buildings that were previously considered not desirable enough to become cooperatives (such as walk-ups in the heart of the West Village), I helped transform the landscape of New York City real estate and TEI from a one-man operation to a diversified investment, development, asset and property management, licensed real estate brokerage, and alternative energy company with a real estate portfolio of approximately 43.5 million square feet of residential, industrial, office, and retail property. Constantly innovating and finding partners with new ideas is how we grew to own 334 properties across 37 states, five Canadian provinces, Anguilla, England, Germany, Italy, and the Netherlands. No deal is too small or obscure for me to consider, since you never know where things will lead.

When Jerry approached me about 1000M, he wasn't a stranger. We had a history, having already worked on a different development that was also in Chicago.

JERRY KARLIK

Principal, JK Equities

I had been building in the South Loop for many years when the recession hit in 2008. We had developed more than seven different properties, but now we weren't doing anything because there was nothing to buy. In 2011 a property came on the market—125 East Twenty-First Street, a former warehouse on South Indiana Avenue, once an industrial area known as Motor Row. Through our efforts, the four-story building, listed on the National Register of Historic Places, was the kind of redbrick factory building with timber and masonry that people now like to live in. In general, they are great for conversions. But another developer who planned to turn the warehouse into condos was derailed by the market crash. The property was in foreclosure and the developer filed for personal bankruptcy.

Our $3.8 million bid for the building was rejected. The bank went to contract with somebody else, but the deal fell apart. We offered the same amount and again were rejected. When the bank returned for a third time, we *lowered* our bid to $3.6 million—and it was accepted. After putting down $300,000, we were on the hunt for equity.

There was no money available. Banks weren't lending. They were merging, failing, and being taken over by the FDIC. We hired a local Chicago broker to find us some money, but we couldn't get anybody to bite. Nobody understood it. The methadone clinic across the street didn't help. After a prospective investor was put off by the clinic, we told the broker to bring people between 10 a.m. and 3 p.m. to avoid the morning and afternoon drill of folks lining up for their medicine.

Francis, who learned of the project through his broker, wasn't bothered at all by the methadone clinic when he came to Chicago to look at the property. In fact, one of the very first real estate deals that he made, when he was still a teenager, was a rental to a methadone clinic!

A tough negotiator, he came in for 75 percent of the deal to redevelop this property into fifty-nine apartments and upward of eight thousand square feet of retail. Jordan oversaw the construction, which cost $8 million. Our concept was to rent the apartments rather than sell them since there was no clear condominium market at the time. That's exactly what we did when the newly minted Vesta Lofts were completed in 2013.

We refinanced the project using a new $10.6 million mortgage in 2014, getting out most of our equity. It also provided good cash flow for the next six years as demand for apartments in the neighborhood rose. When we sold the building in 2020, the $27 million price represented a 72 percent increase in value in less than six years.

It was a very good deal.

FRANCIS GREENBURGER: We had just completed Vesta Lofts with JK Equities when they approached us about the South Michigan property. As happy as I was with the condo conversion on Indiana Avenue, Jerry's latest proposal was too speculative for me. The idea was to purchase 1006 South Michigan, overpriced by $3 million, in the hopes that somebody would come along and buy 1000 South Michigan next door and our air rights for a tidy profit. I proposed that he explore whether we might buy the land next door, 1000 South Michigan to develop both parcels. This would be ground-up construction of a big building (although

1000 South Michigan Avenue, 2014

at that moment, neither Jerry nor I knew just exactly *how* big it would become). For the right number, I might be interested.

The Karliks learned that 1000 South Michigan was indeed owned by First American Bank, which had foreclosed on the property in 2010 in yet another failed development project. So we moved ahead and put 1006 under contract for a purchase price of $11 million, with the plan to offer the bank $15 million for the empty lot next door. The only problem was that the Karliks couldn't get in contact with the bank that owned it. They tried calling, emailing, and going through brokers. The bank refused to talk to them.

With the impending closing of the neighboring office building, the Karliks were getting desperate. In a Google Earth image of the site, they saw a billboard on the lot for a foreclosure auction with the contact information for the broker, Rick Levin. Jerry had worked with Rick before and called him immediately. Rick, who had a relationship with the bank, agreed to connect the Karliks. That's how Jerry found himself traveling to Chicago on a snowy day in 2014 to make a deal at the twenty-fifth hour.

The history of 1000 South Michigan's value mimicked the roller-coaster ride taken by so many other properties throughout Chicago and the rest of the country over the prior decade. Prices had gone up and up during Chicago's condo boom of the early 2000s, only to plummet catastrophically during the recession. When Jerry offered First American $15 million, it was far below the $25.3 million the bank was owed for its original loan on the site. Still, the $17.2 million we agreed to pay was better than nothing. As the bank's chairman, Thomas Wells, told *Crain's* at the time of the sale, "Our business is lending money, not taking back properties."

At that price, the land on South Michigan worked out to $20 per square foot. My team is used to working in New York, where prices of land can easily run $500, even $1,000, a square foot. A tiny land cost, particularly one with a location as good as this, is the ideal opportunity for a profitable development project. Generally, in New York City, the land is about 30 percent of the budget, construction is 50 percent, and soft design costs are 20 percent. The land across from Grant Park represented only about 6 percent of the budget. After purchasing 1000 South Michigan for $17 million, the appraisal came in for $32 million. Long story short, we thought that this was a lot of property for a reasonable price.

Still, there were issues with the property—complicated issues. Perhaps the most critical was that the site was the southernmost edge of a fifteen-block historic district along Michigan Avenue. What could we build within a historic district? There would be meetings with the alderman and the city's zoning department to find out, but when politics is involved nothing is easy or assured. Chicago was trying to rebound from a major economic slump, and we understood its leaders wanted to encourage development. Still, it was far from clear that there would be the political appetite for a major tower. All development projects come with some uncertainty, but this level meant we had to hedge our bets. So we entered into an option agreement with the bank for the empty lot, even though we had already bought the office building next door, hoping it would all work out. If it turned out that we didn't get the right approvals, then we would wind up owning an office building for which we paid $3 million too much. I was okay with taking that risk, because we were the ones in the position to make the project work, as opposed to waiting for another buyer to come along God knows when or not at all. The bank was so anxious to sell, it agreed to the terms and we started down the path of unknowns to build a tower in downtown Chicago before the clock ran out on our option.

JERRY KARLIK: There's an old expression about boats that I think is also true of real estate. The two best times in the real estate business are when you buy the property and then when you sell it. Everything in between is just a mess.

Chapter 2

TOWERING PLANS, 2015

An Architect Designs a Soaring Seventy-Four-Story Tower On the Shores of Lake Michigan

FRANCIS GREENBURGER: I arrived at the Jewelers Building in February 2015 for a meeting with Helmut Jahn. The ornate turrets, cupola, and terra-cotta edifice of the historic building facing the Chicago River provided a stark juxtaposition to the minimalism of the starchitect's third-floor office.

Soon after we came to an agreement with the bank to option 1000 South Michigan, Time Equities director of development, Rob Singer, headed up a team to figure out just what we could do with the property. Rob is the person who truly manages the project and makes it happen; I'm the granddaddy. His role is bigger than mine. Immediately it became clear to him—and me—that this was an opportunity to build another skyscraper.

Time Equities was still finishing up construction on 50 West, a sixty-four-story, 778-foot residential condo tower overlooking the Hudson River at the southern tip of Manhattan. The glorious building, also designed by Helmut and managed by Rob, had its own harrowing journey. We had just poured the concrete for its massive foundation in 2008 when the global recession brought the entire project to a screeching and unexpected halt. At that time, I had to decide whether to abandon the building or ride out the financial crisis with a massive tarp and an even bigger monthly mortgage payment on the property until the world recovered. I have a strong stomach for risk, but this had mine in knots. However, we rode it out, and by the time I was in Chicago meeting with Helmut on this new project, sales for apartments in 50 West were robust, and it was scheduled for completion in 2016.

An initial discussion with Third Ward Alderman Pat Dowell, whose district encompassed the site, was positive. If we brought the city a reasonable proposal, even one of substantial size, it would be considered. With the political winds at our back, we hosted a design contest and received proposals from top architects from across the globe.

HELMUT JAHN

1000M Architect

Mankind has a fascination with big towers: They have been built as monuments for winning wars and have been built by corporations. They symbolize accomplishment. People now will pay a lot to live at the top of a building. I'm not saying that it is not nice up there—it's different. Maybe I'm biased because I live on the seventh floor of a building right on Lincoln Park, which has only eleven floors. Just above the trees, I can see the lake and the park. If I were a floor lower, I would only see the trees. A floor above, probably only the lake.

FRANCIS GREENBURGER: Despite the fierce competition, Helmut came out the winner. The decision was not based solely on our previous track record. Before I started working with Helmut on 50 West, I had been warned that he could be "difficult." I'm not sure I would expect anything less from the German-born architect who moved to Chicago in 1966 to study under the great modernist Ludwig Mies van der Rohe at the Illinois Institute of Technology. Although he never graduated, he

Helmut Jahn

found a permanent home in Chicago, designing some of the city's most iconic buildings, from the Xerox Center—with its curved-aluminum and glass exterior that was a marvel back in 1978—to the United Airlines terminal "of tomorrow" at O'Hare airport, famous for its walkway filled with a neon light sculpture that moves to music.

Helmut's works—such as the University of Chicago's Mansueto Library where the books, kept in subbasements, rise through an automated retrieval system to be enjoyed in the glass-domed reading room—were audacious, and so was he. An avid sportsman who drove a Porsche Carrera, he was a competitive sailboat racer. He kept a model of his sloop Flash Gordon, one of several boats he owned, in the window of his office.

"Difficult," however, was not my experience with Helmut. Like all successful architects, Helmut had a strong ego and clear vision. But he was an extraordinarily keen listener while we worked on 50 West. He made it his practice to present his evolving ideas to me constantly while soliciting additional feedback. When he was in the middle of a competition, he was in my office every week asking me to look at new sketches and offer my opinion. The next week he would be back, saying, "I listened to you and I came back with this. How do you think this looks?" By the end of the 1000M competition, Helmut must have come up with at least ten different schemes and convinced the Karliks as well that he was the best architect for the job.

In Helmut, I found someone relentless in his desire to improve his work. Just like the sailboat racer he was, he was always willing to reset and trim the sails of his boat, in search of a perfect connection with the wind.

PHILIP CASTILLO

Executive Vice President, JAHN

I was born in Highland Park, a suburb of Chicago, and studied architecture in the city at the Illinois Institute of Technology. In 1979, I was working for an architect who had been one of my professors at IIT when Helmut finished designing the Xerox Center, now called 55 West Monroe. I thought the forty-five-story office tower in the heart of the Chicago Loop, which would go on to define the contemporary curtain wall, was one of the most important new designs at the time. I decided I wanted to go work for Helmut, and, with the exception of a brief period of a few years, I've been here with him ever since.

I've worked on some of the most important projects in the office, like the Sony Center in Berlin, Suvarnabhumi Airport in Bangkok, 50 West Street, and 1000M. I like to say that if you took one picture of every building starting with the Michigan City Public Library, which Helmut designed in 1974, and lined them up, they would tell a cohesive story. Even though they might change, there are still pieces that are a reference from other jobs that continue to inform. No building is a duplication. We resurrect them in different ways, shapes, or forms.

It's a challenge every day, but I love the challenge. We've had such a long and close working relationship because I understand

Philip Castillo

where Helmut wants to go. He has a certain level of trust that I know when to be pragmatic and how to get a job done.

I'll be honest, though, I can never keep up with him. He's the consummate twenty-four-hour architect. I get emails at all hours of the night from him. At 5:30 a.m. when I wake up in the morning, I sit at the kitchen table and see what's come over the night from everywhere and anywhere—including Helmut.

FRANCIS GREENBURGER: I shouldn't have been surprised that February day at Helmut's office when walking into the conference room I was greeted with drawings of forty-plus schemes covering the walls. On closer inspection I could see that each "final drawing" had up to ten different drawings pinned underneath its final iteration. For our meeting on the "massing of the building"—an architectural term referring to the general shape and size of the structure or how it will look as a form against the skyline—Helmut and his team had prepared multiple configurations and variations for each idea. As one looked from drawing to drawing, one could see Helmut pushing himself to reinvent his ideas, innovate, and find another, better way to express them. While there were multiple schemes and approaches, each drawing clearly conveyed his vision. Helmut had a phrase he muttered often: "We can always improve."

It took probably fifty different models in Helmut's office, many hundred more drawings, and several months, but eventually we finally settled on one scheme for the massing. To maximize lake and park views and the total available square footage for apartments, the tower was essentially a 285-foot high base (to mimic the building immediately to its north) with three-stacked cubes that cantilevered out at an angle. Structural glass instead of metal was a way of reducing the boundary between the interior of the condos—including everything from studios to a full-floor, 10,500-square-foot penthouse—and the great vast expanse of Lake Michigan and its shoreline.

On October 29, 2015, almost four months after we came to an agreement with the bank to buy 1000 South Michigan, we presented our plan for a 1,300-foot tower at a packed public meeting held less than a block away from the site. We were asking to go a whopping 575 feet over the zoning limit for the historic Michigan Avenue Boulevard district.

The property came with about 900,000 square feet of buildable footage. But with a height limit of 425 feet, the same volume configured differently would give us a very fat, wide, and short building. And that's no longer what people want residential towers to be. Low, wide buildings are efficient but not sexy. They result in deep, dark apartments. People want to be higher in the air where the views are better. Well, most people. But would the zoning board approve of such a drastic change to the neighborhood's rules around height?

Navigating the rough waters of real estate regulations for 1000M was Jack George, a Chicago native who has practiced law for half a century in the city he has lived in all his life. He started out his career in government, first as an assistant attorney general for the State of Illinois and then as an assistant corporation counsel for the City of Chicago, before going into private practice. Concentrating on land use

Original Competition Design reduced to 832 ft.

and development work for the last forty years, Jack has worked on many major Chicago developments, including Central Station, an eighty-acre residential development project on industrial land and rail yards in the South Loop that began in 1988 and took twelve years to complete. With his deep knowledge of land use and zoning rules, Jack was invaluable in helping make many buildings along Lake Shore Drive become a reality.

JACK GEORGE

Counsel, Akerman LLP

Over the years I have, in my own small way, helped in shaping the skyline of Chicago. 1000M was one more building—but a very important building because of Helmut. It was a great privilege for me.

We had to appear before the landmarks division in the Department of Planning and Development and make our case about why the height restriction should not be enforced. Our argument was that other tall buildings built around the site made the old height restrictions not only obsolete but out of place. For example, the Columbian, a forty-six-floor tower built in 2007, was only one block away on South Michigan Avenue. We believed, when considering the surrounding landscape, there was a need to have a building of the height that Helmut proposed.

From my first project on, I have always enjoyed this work. During my early years in private practice, I had been doing other legal assignments when a client wanted to do a development in a western suburb of Chicago. From that point on I was hooked, and I have been involved in hundreds of cases. It's rewarding to see these developments, which are so important to the community, approved by various municipalities. I enjoy working with the greatest architects in the world and the specialists in building a case—the most challenging part of the job.

When a client comes in with a project, it is my responsibility to say, as best I can, whether this proposed development will meet the criteria of the numerous regulations of various agencies, including transportation, landscaping, and the Mayor's Office for People with Physical Disabilities, to make sure the building meets the accessibility and adaptability requirements of the local and federal government. If in my opinion the project is viable, I have to bring together the whole team and formulate a strategy for getting it approved.

Then comes the back-and-forth between the client, various agencies, and affected community groups. While ultimately it's my job to advocate on behalf of the developer, I have to listen not only to the client paying my fee but also to the groups and agencies I appear before to make the case. I have to be knowledgeable about the regulations *and* be somebody who listens.

When meeting with community members, you can't just say, "Jeepers, they're crazy!" They are not there to cause you a problem; they are raising legitimate concerns they want to address. I always try to put myself in the position of a homeowner who lives by the site. If I were an owner, what would I be concerned about? In the case of 1000M, we had a residential building right next door. If I lived there I would worry about the design of the building, how the developers were going to handle the construction process, and how it will affect traffic.

Here the building was beautiful. No one could criticize the design. But neighbors were concerned with the meat and potatoes stuff. How is it going to affect my lifestyle? If I put myself in their shoes, I am prepared when I go to the meeting to give good responses because my job is to listen—and respond.

FRANCIS GREENBURGER: Chicago in particular is pro-development, even given the anti-high-rise members of the community who packed the October 15 meeting. It is definitely a lot friendlier than anything I've experienced in my decades working in New York real estate. My first inkling of this came much later at an event the sales staff put together for potential buyers at 1000M's sales center. I was one of two speakers that night, but I didn't know who the other speaker was. Soon enough, I discovered she was the buildings commissioner for the City of Chicago! In New York, the buildings commissioner is like God in that no mere developer is ever able to see them—let alone have a conversation with them, at least in my experience.

After her presentation, Chicago's commissioner, Judy Frydland, chatted with me, in a friendly way. That wasn't all; she also gave me her cell phone number and told me to call if I had any problems. I have never called the number, but I did appreciate the big difference in attitude from my home city, New York.

Despite Chicago's positive outlook on high-density development, our contract to buy the land that 1000M would be built on was contingent on our getting through the entitlement process. Because we knew we wanted to do something the in-place zoning regulations didn't allow for, we made a deal with the seller that we wouldn't close until we got permission from the city to erect the kind of tower we wanted to build on the site.

Luckily for all parties, the city supported us through the entire legal process and ultimately found an ingenious solution to the height-limitation issue: move the boundaries of the historic district.

The decision wasn't arbitrary; the department's priority was having a continuous street wall along Michigan Avenue. When we bought the site, the continuous street wall didn't exist because of the vacant parking lot in the middle of the block. Helmut's design respected the continuous street-wall design and intent and therefore was seen as a desirable improvement.

Our site consisted of a parking lot with eight-story and three-story nondescript buildings in succession—only two blocks from the border of the historic district that runs about a mile. For these reasons, the planning department made the argument to the city council and mayor that our block shouldn't have been part of the historic designation in the first place. To the planners, it was more important architecturally to extend the street wall than to protect these nondescript buildings, so they proposed moving the border from East Roosevelt Road to Eleventh Street.

DANIEL KLAIBER

Coordinating Planner, The City of Chicago
Department of Planning and Development

Growing up in suburban Detroit, I was always fascinated with cities. Becoming a city planner was an answer to that early curiosity about the forces and reasons behind this big city getting built—and maybe being able to fix parts of it. I was definitely one of those kids who kind of wanted to save the world, or at least be part of a community trying to make a difference.

For my first ten years at Chicago's Department of Planning, I worked on neighborhood redevelopment. My coworkers and I were liaisons between the aldermen and the mayor, shepherding different projects on a smaller scale throughout the city. By the end of Richard M. Daley's mayoral term, I had transitioned to citywide planning where I eventually worked on a big retail study and commuter train-station plan. My role changed during Rahm Emanuel's administration. Plus, the market was really heating up, so at the tail end of December 2013 I was moved to planned developments. As a unit of zoning, we were the quarterbacks for large-scale development projects.

The job of the professionals in our department is balancing the public interest with that of the developer while adhering to the zoning code. We crunch the numbers and look at the plans to figure out the parking, height, design, number of units, and everything else important to the project. There are dozens of questions, from the number of affordable units to bike parking.

For 1000M, height was the big issue, and how a tower of its size would look within the landmark district. The theory was that the height would be okay because when we looked at the southern end of Grant Park, there were taller buildings outside of Michigan Avenue that framed the whole park. Following this logic, Michigan Avenue at the southern end would ramp up to the higher end of buildings on

Roosevelt. In fact, I had been working on another large development just a couple blocks north of 1000M. The fifty-six-story apartment tower, then called Essex on the Park, was smaller than 1000M but still quite tall.

My work, embodied in my title, is a lot of coordination—among attorneys, architects, the community, elected officials, and my own department. We end up spending the most time with the attorneys, exchanging comments and letters. But in some cases, we also meet directly with the architect, which we did with Mr. Jahn. I remember that because Mr. Jahn makes quite an impression. I certainly will never forget his presence at the Chicago Plan Commission.

The ultimate goal of all the work—all the calculations, documents, and exchanges—is a trip to the plan commission. After the plan commissioners weigh in, the proposed development has to go through a city council vote. But this is the big, showy hearing with a PowerPoint full of graphics that hopefully make a big splash. It's also where the public really gets to see the development plans in full and provide comments.

Because these are not City of Chicago projects but private developments, the project managers from planning would lead off the hearing and then typically let the development team take everyone through the entire project. In this case, however, our commissioner wanted us to present the whole thing. I would be lying if I said I wasn't a little nervous. This was definitely one of the biggest hearings I had ever participated in. But I had my notes, which I had studied, and the flashy PowerPoint; I was ready. Helmut Jahn came in and took his place right next to me. I glanced at him, and he's wearing cool jeans and shoes with no socks! "Oh, wow," I thought, "He's just like Mick Jagger."

In the *not*-cool suit I wore to all my hearings, I went through the entire presentation: the base, the parking, the floor-to-area ration, all the boring stuff. Then I got to the tower itself and attempted to describe it: "It twists, and it does this and that . . ."

"What are you doing?" I thought to myself.

I'm a city planner, not an architect, let alone a starchitect.

When I finished the presentation, there were crickets. Of course, the commission didn't want to hear from Dan Klaiber. They wanted to hear from Helmut Jahn, and I didn't blame them.

My commissioner addressed Mr. Jahn directly and asked if he would like to address any comments directly to the commission about his design.

Mr. Jahn stood up in front of the commission and said, "Dan, I think you presented the building very well from what it is. Let me maybe make a couple of remarks." It was very gracious of him.

When the meeting was done, I asked Jack George if he wouldn't mind asking Mr. Jahn to sign my staff report for me. There were a lot of people who wanted to talk to him, including members of the press. But Mr. George returned just a bit later with my staff report signed by Mr. Jahn.

Whatever the city or neighborhood, at the end of the day, a city planner is charged with acting on behalf of the public interest. That doesn't mean you are always going to please the public. It sounds kind of simplistic, but you are trying to make the city better. And for the most part, I think we *do* do that.

FRANCIS GREENBURGER: In New York, this proposal would have resulted in dozens of tabloid headlines and people picketing. New York's city council members would jump into a fire before they moved a historic district to allow for a new development—no matter how much it made sense. Yes, Chicago was a very different place. The city voted to approve the proposal in a transparent process covered by the *Chicago Tribune*. Once they made the decision that we were not in the historic district, it paved the way to get approval for the building plans.

PHILIP CASTILLO: We wanted the building to be a thousand feet tall. But while I talked to someone from the zoning department in the lobby of city hall, he made it clear that wasn't going to happen. So what did he think they were going to approve?

"Probably six hundred feet," he said.

"What about nine hundred sixty feet?" I countered.

That was unlikely, he said. But how the board came up with the final height limit of 832, I have no idea.

ROB SINGER

Director of Development, Time Equities

When the officials from the planning and zoning departments informed us that they had determined our maximum building height of 832 feet, we said, "OK, let's just round it up to 850?" They looked at us and said, "What don't you understand about 832 feet?" We got it—832 feet. Random, but we'd make it work.

JACK GEORGE: We went ahead and argued for a zoning increase of 450 feet to 832 feet. They agreed the height restriction should no longer be enforced. After we got that approved, we were able to move forward with our design. It was a major victory.

FRANCIS GREENBURGER: We finalized the plans and approval for what the *Chicago Tribune* described as "the tallest tower in the iconic row of skyscrapers on South Michigan Avenue . . . [and] the tallest Jahn-designed building in Chicago, where the architect is based."

I was in Nicaragua for Christmas, because my youngest daughter, Claire, is an avid surfer. I was showing everyone pictures of 1000M to see what they thought when Claire's childhood friend Myrto, vacationing with us, said, "It looks like it's going to fall over."

"What?"

Early facade sketches by Helmut Jahn

"The top is bigger than the bottom, so it looks like it's going to topple over."

For the next few days I reflected on Nikki's comment until I realized she was right—and that nobody was going to buy an apartment in a building that looked like it was about to fall over. On all levels, you want to be secure in your home.

I picked up the phone and called Helmut.

"You're going to kill me," I said. "You have to go back to the drawing board."

HELMUT JAHN: I was skiing in Utah with some friends when Francis called to tell me that he couldn't sell apartments in a building that appears to be falling.

There was a reason it leaned to one side in the front and became more slender in the back. This way people, who lived in the back in apartments over the parking lot, could also look toward the lake because the building flaps out. At the same time, giving a historic gesture to the street wall was a big part of the city approving the building. The lower portion of the facade was more solid. As it transitioned off the street wall, the tower became lighter with less aluminum and more glass. So the building had respect for its context but became a very free shape as it rose to the sky.

There would have been a time when I would have been really mad about that call. Instead, on New Year's Day, I called Phil during the car ride from Deer Valley to the airport: "We have to change the design." I didn't have the right paper, but I took whatever I could get and began to sketch. The first one I made was on a legal notepad. It was crude. Two days later, I made another sketch. Then another one on January 4, my birthday.

PHIL CASTILLO: We knew we had to change the building. If Francis doesn't like the design, he isn't the kind of guy you're going to convince otherwise. He cares about the architecture, but the overriding point for him is always what makes someone buy into the building. That includes not only looking at it from the outside but the experience of being inside it.

I remember while deciding on the windows for 50 West, we were going outside to look at the glass. Francis said, "We don't have to go outside. The guy buying the apartment doesn't care what the glass looks like on the outside. He cares about what he sees through the glass when he's looking outside."

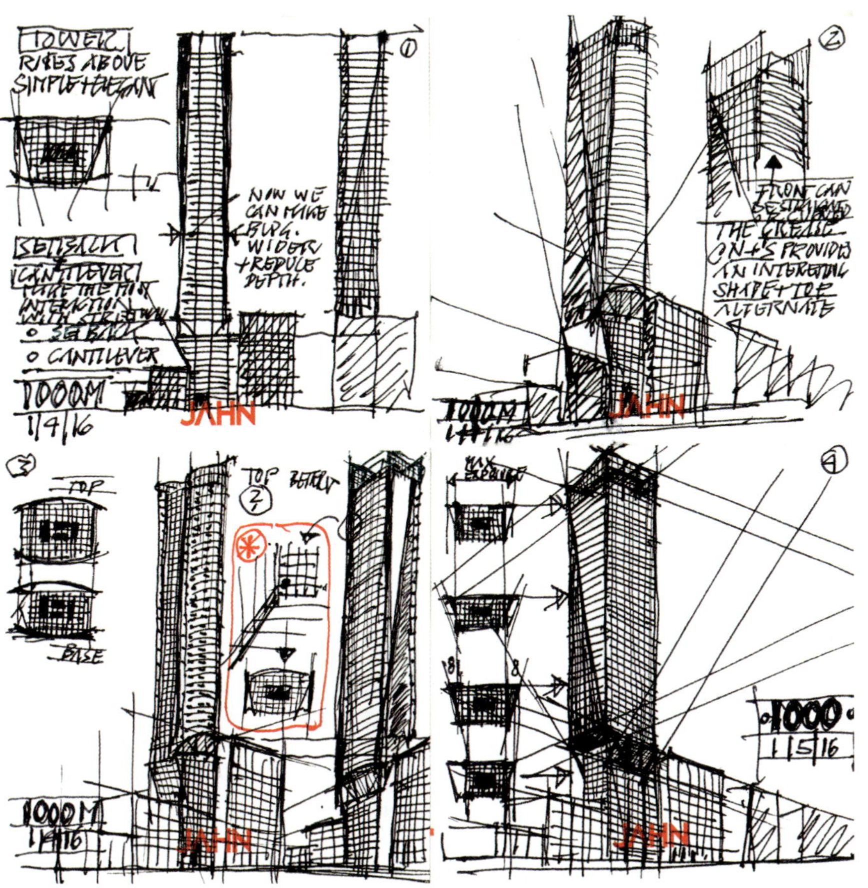

Early massing studies by Helmut Jahn

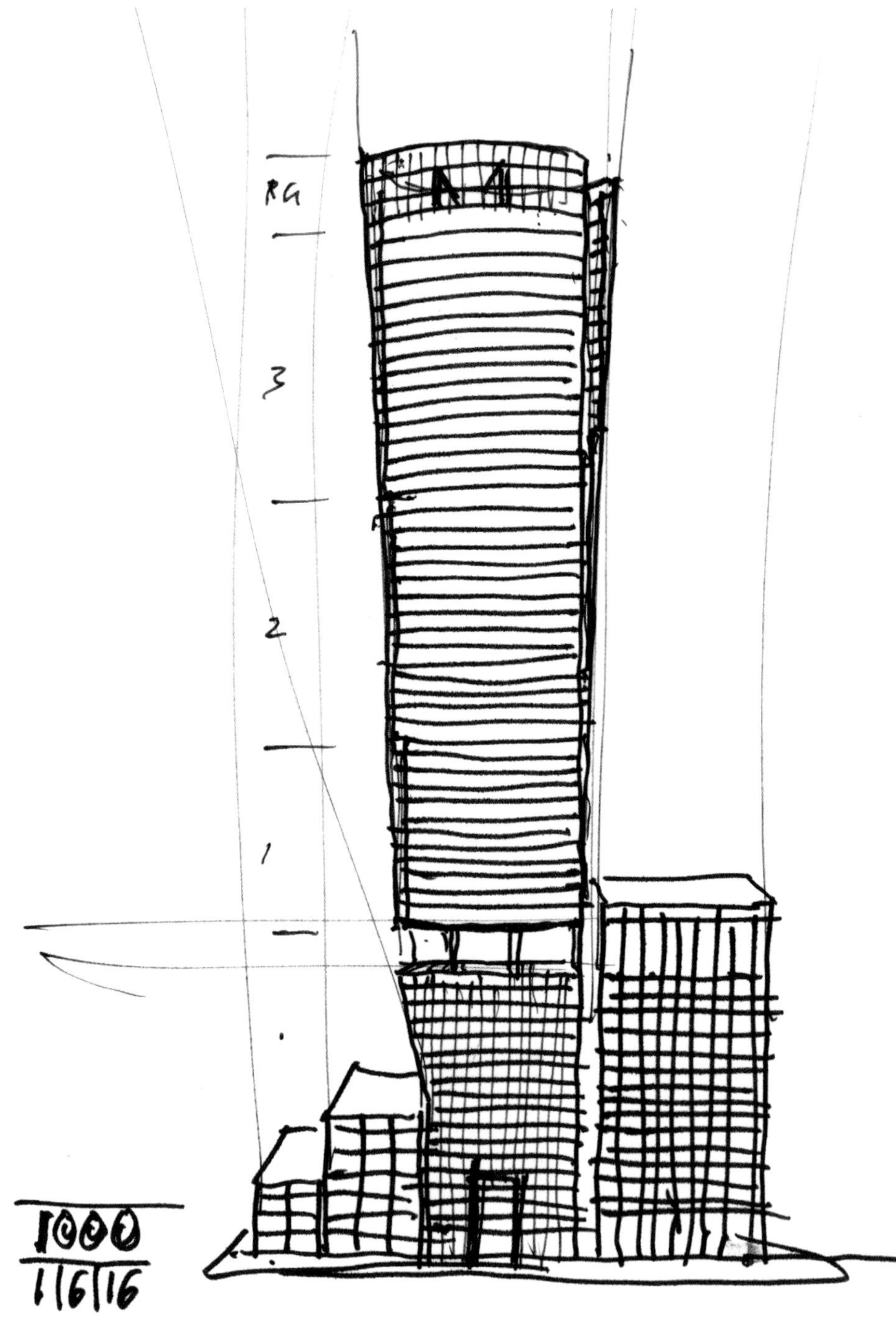
RG
3
2
1
1000
1|6|16

Helmut Jahn sketches "evolution of design"

The transparency of the glass wasn't the only detail he paid attention to when it came to the windows. Our original design for 50 West had windows that extended all the way down to the floor. Francis insisted on a sill that went a little more than a foot up from the ground, because he didn't believe people would feel comfortable with a floor that appeared to drop off like a cliff's ledge. Helmut and I didn't believe that, but there was no reason to argue with him.

FRANCIS GREENBURGER: Not even a month after Helmut's sketches for the new design, we presented the planning board with a new seventy-four-story tower, the tallest in his career. Gradually widening at the northeast and southeast corners, 1000M bloomed out of its solid rectangle base to an astonishing parallelogram that afforded all sides of the tower expansive views from floor-to-ceiling glass panels. Just as compelling as this feat of architectural imagination was that the building's height fit into the amended guidelines of the historic district set by Chicago's landmarks commission. In April, the Chicago Plan Commission approved the proposal and city council not long after.

ROB SINGER: After various design iterations, redesigns, ups and downs, the tower design was efficient, practical, and beautiful. It had interesting architectural moves that gave it a unique shape and responded well to its context and environment. At the eighteenth floor, 1000M begins to slope out over the roof of the neighboring building to the south, 1006 South Michigan, the small office building we also purchased in the original deal. Above the twenty-first floor, northeast and southwest corners start to gradually flare out as the tower rises while the northwest and southeast corners stay fixed—giving it the appearance of a cobra head and increasing the floor plate sizes on higher, more valuable levels.

Those flaring corners in the southwest and northeast corners also improve views of the water and downtown from the apartments in those unit lines. It was an inspirational meeting of form and function, which is the "gold" Francis is always searching for.

HELMUT JAHN: The interesting part is that the building's footprint is essentially a rectangle. The facade of the first forty floors has a respect to the context of the street wall on either side of it, which is more solid. A big cut marks the transition off the street wall as the tower rises. Past the fortieth floor, it turns into a parallelogram with totally freestanding walls. This is the endpoint of the sheer walls—walls where the core is connected to the supporting columns. The aluminum and glass of the parallelogram also appears less solid than the lower portion; it becomes a very free shape in the sky.

There are other towers with sheer walls but not to this extent. I think this one is probably a little more aggressive than others. I have no problem with the way the building looks like right now. This is definitely more pleasing than the first design. I thought, it's going to be a very good building if we get the right curtain wall and the right materials.

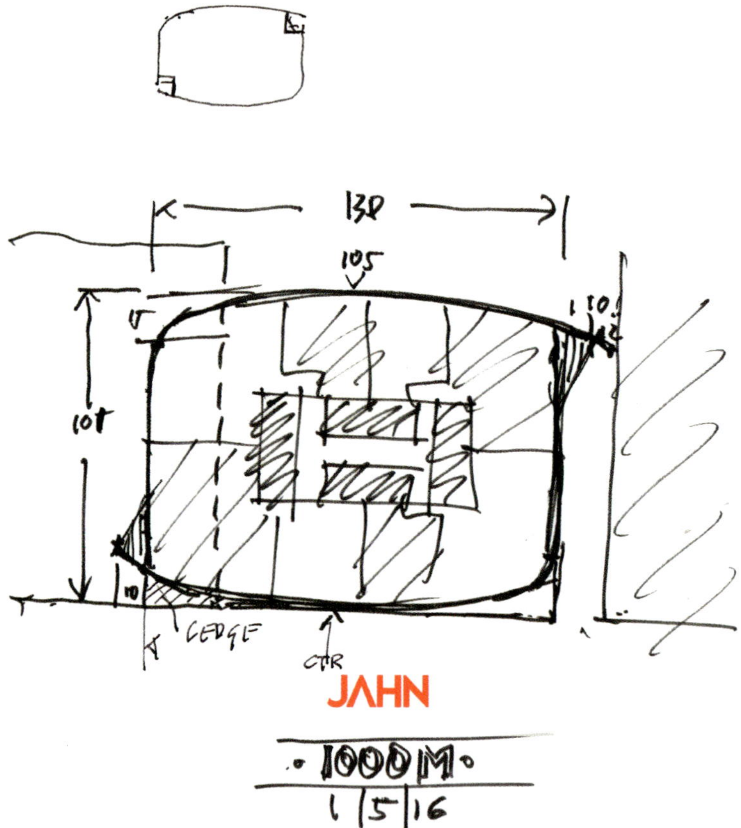

Floor plate concept by Helmut Jahn

Chapter 3

BUILDING A DREAM, 2017

Turning a One-Million-Square-Foot Blank Space Into a Place Called Home

FRANCIS GREENBURGER: We put $30 million into 1000M by fall 2017 and were still a year out from the projected groundbreaking. The price of the land was $23 million, and the rest could roughly be described as architectural and design fees and marketing costs.

For the interior design, we hired Kara Mann, a superstar in the industry known for her sophisticated, edgy projects that cull from a wide and unexpected range of materials. Her unique mix of hip and formal made her the go-to designer for tastemakers such as Gwyneth Paltrow, who collaborated with Kara for a pop-up shop of her lifestyle brand, Goop, in Chicago.

KARA MANN

Founder and Creative Director, Kara Mann Design

I love Chicago. It's my hometown; I grew up in Evanston, so to be able to put a mark on the city and work with Chicago superstar Helmut Jahn was exciting.

Kara Mann

With any project, I like to start with its geographical context and how that plays a role in what I'm designing. It's also important to think about the style of architecture that we're designing for, because it has to feel harmonious. What's so interesting about Chicago is the proximity of the city to the lake, and the way they collide into one another. Our pitch for the design of 1000M highlighted that juxtaposition and the surrounding park system.

Then you need to learn about the client: Who is going to be living in this space? Who would live in that area of the city? What's their mindset? The developers were interested in us being integral to the team developing the layouts, working with the sales team from the very beginning to come up with the unit mix and amenity spaces.

We start very conceptually—very big picture. As a firm of both designers and architects, we start with an understanding of the space's architecture. Then we layer in the materials, furnishing, and mood.

I studied art and worked as a stylist in the fashion industry before becoming a designer. Interiors are my third life, so I combine my past into the language of the spaces we create. We pull references from everywhere: We pull historic referencing; we pull current referencing; we pull fashion; we pull art. Francis and I were definitely aligned on the importance of art in an interior. I never feel like anything's finished until there's art in the space. He's got a rich language and history in design, so that made the collaboration fun.

Amenity spaces are now key to selling a building. While it's fun to dream up interesting new offerings, the latest and greatest, we teamed up with the developers, architects, and sales team to figure out what would actually be desirable for potential buyers. The challenge of the amenity spaces for us, the designers, is how to program to square footage. We had to make these spaces interesting, inviting, and usable—each with its own personality but that integrated a cohesive story that draws people through the building. It's a balancing act to make large-scale spaces feel intimate and personalized.

FRANCIS GREENBURGER: For 1000M, her first ground-up condo tower, Kara and about twelve people from her team started the programming phase of the project, which began with senior designers creating the vibe. In drawing inspiration for the look and feel of the various spaces, they create a very conceptual mood board. The images don't have to be furniture or design materials. They can be anything—even a woman's dress.

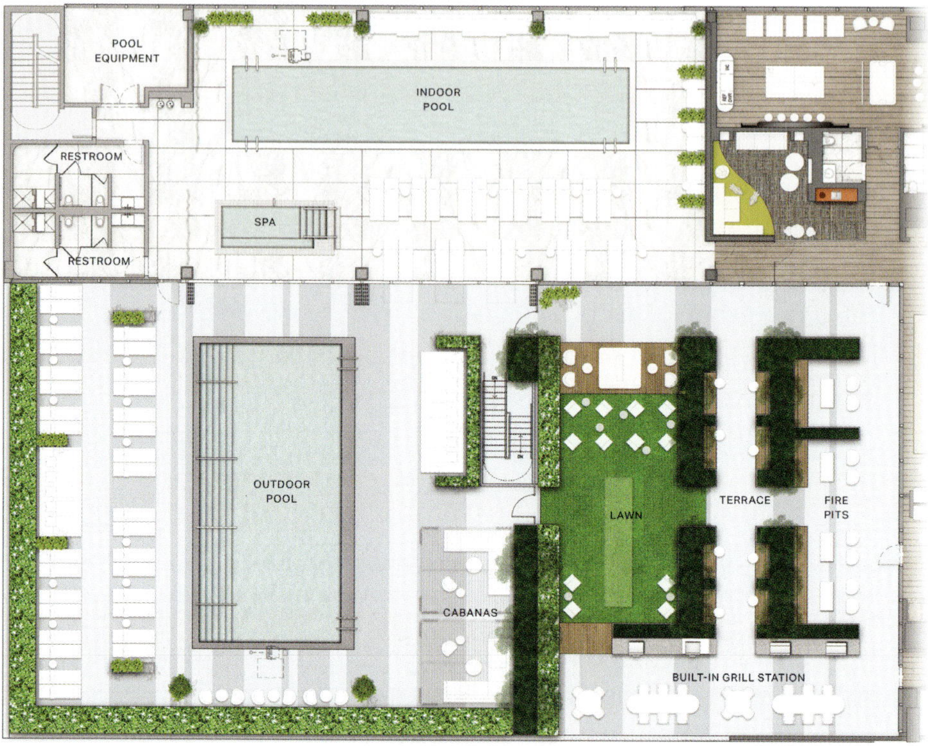
POOL
EQUIPMENT
INDOOR
POOL
RESTROOM
SPA
RESTROOM
OUTDOOR
POOL
LAWN
TERRACE
FIRE
PITS
CABANAS
BUILT-IN GRILL STATION

PLAN

01
02
03
04
05
06
07
08
09
10
11
12
13
14
15
16
17
18
19
20

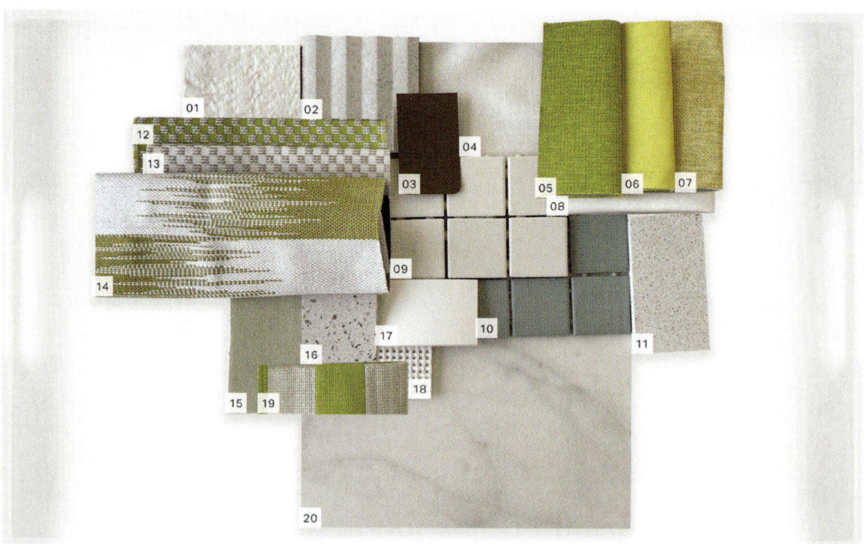

Pool plan and palette

Exterior pool

Winter garden design concept and early materials palette

73rd floor with foliage

LAURA LEE MCALLISTER

Director of Architecture, Kara Mann Design

The conceptual presentations create a voice and a vibe for the interior of the building. We use conceptual precedent imagery to convey a mood or a design theory. If the idea board for a library includes something like a cropped picture of a flowing dress, the client might respond, "What are you saying with this?" To us, the designers, it means a lot. Maybe the ridges in her dress will become the wall covering. We just need some place to start.

Once the client or developer approves of the general scheme, our team gets into drawing every elevation from the ceiling to the floor. We do 3D perspectives to make sure that we're shaping the room correctly. Our furniture team layers in furniture, fabrics, and finishes—rugs, lighting, and window treatments. We design the entire room, including furniture, at one time to convey the complete design to the client. We are consistently working with the architect to confirm that our design works with the overall building.

I'm actually a trained architect, not an interior designer. Since I grew up on the coast of Alabama, I studied architecture at Auburn University before moving to Chicago about ten years ago. I worked for a couple architecture firms until I found my way to Kara, where I have been for almost six years now.

My architecture brain needs to finish a room—I like the idea of it all being thought through and designed at one time, down to the furniture and accessories. While my training was more about the exterior, I'm more passionate about the interior layout and the layers of design and how they work in tandem with the architecture. What about the sofa? How is the window treatment going to work? At the same time, my training allows me to bring a surface-level knowledge of the work that Phil and Helmut are doing to help our offices work well together. I can bridge the conversation between the architecture and interiors teams, finding the right balance.

To present the next level of the design, we create trays with physical samples—swatches of fabric, pieces of stone, wood, metals for fixtures—that we support with drawings, renderings, images, whatever we think will tell the story. Once we get the sign off from the client, we produce construction documents where every dimension and material is called out, which we pass along to the architect, who incorporates that into the bigger architectural set of plans for the contractor.

When we present this phase of the interior design to some developers, their attitude is, "Looks great. Move on." Meetings with Francis are always much more collaborative. Because he really cares and thinks about the materials, he is always tweaking things and suggesting new ideas. Before Francis gives his approval, we go back to each piece, perfect it, and make another presentation. Although the process is more work, it's nice that he cares and is so passionate.

FRANCIS GREENBURGER: For the 323 homes in 1000M—ranging from 926-square-foot one bedrooms to 5,491-square-foot penthouses—Kara and her team created three custom color palettes for residents to choose from. One of Kara's signatures is the use of stone, in particular veiny stone, which found its way into the kitchens boasting Wolf and Sub-Zero appliances, bathroom "wet rooms" with large soaking tubs, the two-story Michigan Avenue lobby, and much of the 40,000 square feet of indoor and outdoor amenity space.

Amenities have become a huge part of property development. Gone are the days when a bicycle room and gym sufficed. People want extensive communal spaces with programming and in-house conveniences. Here's some of what we planned for 1000M:

- full-service concierge to help residents arrange for services and plan events
- resident-only dog park and dog spa with washing and drying stations on the lobby level
- state-of-the-art fitness center
- beauty bar with a hair blowout station
- spa with a salt room featuring walls adorned with Himalayan salt, hot and cold plunge pools, and a virtual meditation lounge
- library lounge
- conservatory for musical performances and art exhibitions
- golf simulator
- billiards lounge and game room
- children's room
- sound studio for music practice and production
- chef's catering kitchen
- coworking space and conference room

We designed outdoor spaces on the eleventh floor with formal gardens, walking paths, hammocks, grills and dining tables, and a fifty-two-foot swimming pool with private cabanas. There's something for everyone, but perhaps the most stunning communal feature of 1000M is the seventy-second-floor winter garden. The ten-thousand-square-foot common rooftop terrace at eight-hundred-feet high was meant to be one of the highest in all of Chicago. With nineteen-foot floor-to-ceiling glass walls offering one-eighty views of the lake, park, and skyline, year-round indoor green walls, telescopes for star gazing, and a full-service bar and wine tasting room, the "winter garden" has promised to be one magical place. Now we just had to prove to potential buyers that 1000M was a building special enough to justify the hefty price tag of its condos, which begin at approximately $557,000 for a one-bedroom and run up to $8.5 million for a four-bedroom penthouse.

Despite some major developments that had cropped up in the area, we had to convince people to pay higher prices for a pioneering location within Chicago's condo market, which had a fickle history that was made even more difficult among the many failed projects pre-2008. Meanwhile, condominiums cannot be built without a considerable part of the building presold; banks will not finance a project without proof they can recoup their loan. To persuade buyers to purchase high-priced apartments that don't yet exist requires very good marketing and sales teams. We had to give buyers a sense that 1000M was extraordinary.

Otherwise Incorporated, the Chicago-based branding, strategy, and design agency we hired to helm the marketing, pitched fifty different concepts before settling on the winning campaign. The boutique firm is a husband-and-wife team of writer and business strategist Nancy Lerner and designer David Frej, who met on a project in the late 1980s and, a year later, wed and went into business. Their happy marriage of words and design produced a simple, yet genius campaign for the conundrum of how to advertise that an empty parking lot would one day sprout a monumental piece of architecture: Seek Beauty. Two sleek and simple words conveyed so much allure. In a sizzling way, the phrase encapsulated our aspirations of bringing a building of high quality and design to the less developed end of Michigan Avenue.

NANCY LERNER

Cofounder and Chief Strategist, Otherwise Incorporated

In building brands, everything we do derives from the idea that organizations, enterprises, and people have a fundamental character and value. By illuminating the ways in which our clients have a different way of being in the world, it doesn't become about just selling stuff but something deeper. This really applies to real estate where branding is usually a farce. The passion for architecture and design that goes into a building rarely translates to anything that is a brand. In service of leasing or selling, the building itself is supposed to carry the brand with everything.

Chicago is an interesting place to create the branding for an important new tower. As practical and pragmatic Midwesterners, Chicagoans don't fetishize buildings or specific addresses the way New Yorkers do. At the same time, they are fanatical about architecture. We knew that 1000M was going to be iconic and change the skyline. We had the added ingredient of Helmut, who is known in Chicago. The Thompson Center—the postmodern municipal building with offices of the Illinois state government—is hated or loved, nothing in-between. Helmut hadn't done a major residential project in the city, so this was a moment.

Everything lined up to give us permission to push the envelope and create a fabulous brand for 1000M.

However, the building was going to take four years to build, and selling 323 residences was a lot. To keep the marketing strong throughout the process was going to be an uphill battle. David and I knew we needed a much broader platform to operate from than typical real estate branding concepts, something universal that would provide all kinds of opportunities for creative expression.

Our methodology at Otherwise is based on a process I call "empathic inquiry." Different from the linear process many creative agencies follow, it is a strange and provocative way of working, based on the belief that if I can bring together a group of people with a shared interest and create circumstances where they can get out of their own way and collectively focus on that interest, we can create something together that is magical, remarkable, and fresh.

"Empathic inquiry" is my own invention and goes back to research I did as an undergrad at Barnard for Alice Pawderer-Singer in her multiyear study of twelve-person juries funded by the Russell Sage Foundation. The basic question she set out to investigate was how a group of people come to a unanimous set of decisions in a trial. What happens is that a natural leader emerges and pulls the others along until they finally get to a unanimous verdict. That's an important process for our criminal justice system, but creatively so much gets lost during this process.

Over the course of my career, I have been interested in a different way that is about diffusing power to harness the wisdom of crowds. My philosophy is that when people create together, the end product will always be better than what anyone can do alone. The challenge is making sure the decisions and ideas are truly collective, which can be hard to do when there are powerful leaders that are used to making most of the decisions.

In order for high-functioning people to participate, they have to summon a lot of patience. I have the ability to remain fluid with massive amounts of information for a long time—to the point where many others get uncomfortable. Successful individuals deal with large amounts of complexity but usually linearly. To create a single story, you have to edit and leave stuff behind. I like to keep things open for as long as possible until putting them in order. The outcomes are always wonderful and robust.

In a real estate project with multiple high-testosterone personalities, this doesn't happen on its own. Francis, Rob, the Karliks—any one of them can hold a stage by themselves. But this isn't just about playing nice: it is the opportunity to create something bigger and better—together.

As we started to dig into the work for 1000M, there were a few lines of poetry on my radar from John Keats's "Ode on a Grecian Urn."

Beauty is truth, truth beauty,—that is all
Ye know on earth, and all ye need to know.

This accessible poetic expression seemed like a perfect source to pull in for the project. It hinted at a simple, yet universal idea that part of the human experience is a desire for beauty. This higher purpose is what separates us from other living things.

Through the group process we created the 1000M moniker Seek Beauty. This message, threaded throughout the marketing campaign, was more than a command to prospective buyers. We had been inspired by the building's "forever views." The tower's location on the lake offers a vista that will never be altered, much like the timeless search for beauty.

The work we did together was so gratifying, partially because this building is an amazing contribution to the skyline of a city that cares deeply about its architecture.

FRANCIS GREENBURGER: In late March 2017, the message to Seek Beauty appeared cryptically along the sides of public buses and a one-hundred-foot blue banner that ran the length of the building site. The intended beckoning effect succeeded. Architecture blogs posted comments. People were curious.

In summer 2017, the sales center opened for early buyers to see what they were actually being asked to purchase. This was no storefront with printed pamphlets. The local real estate marketing firm PJR Design and digital environment developer M1 Interactive won a national award for turning the eight-thousand-square-foot sales gallery into an immersive experience with, among other things, a video wall of twenty-eight commercial grade displays allowing brokers to choose any floor and pan 360 degrees—at day or dusk—to show potential buyers exactly what the view from their window would look like. Today, when you build a building, you build it twice. First you build it digitally. The digital imagery is a whole art form in itself. The renderers work in partnership with the architects, taking the two- and three-dimensional plans and turning them into a full-blown projection of a future reality. The final product looks like high-quality photos of places that actually exist rather than virtual images.

We didn't, however, officially launch the sales for 1000M until October 10, 2017. In a press release we sent out widely, I expressed my excitement about introducing 1000M to the world. "The team behind this project shares a vision of beauty and is committed to exposing that vision inside and out," I stated in the release, which announced the leading Chicago brokerage @properties was helming the sales and marketing of 1000M, as well as that construction was slated to begin late 2018 with expected completion in 2021.

ROB SINGER: The first day we opened sales for 50 West Street, we were overwhelmed with interested buyers. The white-hot market was remarkable. Obviously, we had projections for how sales might turn out, but you never really know until you open the doors and start booking appointments. We were taking ten contracts per week. Even more important than the volume of our sales was the pricing we were achieving, which was around $2,000 per square foot and in line with our projections. Our basis (or break even) was around $1,400 per square foot. This meant we were immediately selling at a significant profit margin. Lenders and investors were pleased, to say the least.

At 1000M, the problem was that our breakeven was around $700 per square foot, and we needed an average of $950 per square foot to create a reasonable profit margin. Meanwhile, the contract sales were averaging around $750 per square foot. We started off with several "teaser" low prices for apartments at the bottom of the building that were actually below our cost. Prices on those units were below our cost basis, so we were losing money selling them. The other units at the higher "real" building prices weren't selling nearly as well.

There were other high-priced condo buildings in Chicago selling for over $1,500 per square foot—far in excess of what we were trying to average—but our location was different. 1000M is south of the river in a less established area. As the months wore on, we continued to try to sell contracts but couldn't lower prices any further without jeopardizing the project's viability.

We continued to churn through winter 2017, believing that it would get better, but it never really did.

Marketing banner

Chapter 4

FINDING THE MONEY, LATE 2017

How to Borrow $343 Million

FRANCIS GREENBURGER: The real adventure of 1000M was securing the construction loan. Generally, banks don't like to put up more than $50 to $75 million for any given condo-development project. Because we were looking for $343 million to erect 1000M, we expected to assemble loans from several banks. When we financed my first skyscraper, 50 West Street, there were six banks involved.

When we went out in the market for bids, people were initially lukewarm. It's always challenging to finance a building based on a pro forma and assumptions of what sales prices, rents, and market absorption will be in two to three years, when the building is completed.

What any construction lender obviously wants to know is, "How do I get my money back once you've built the building?" And for every answer, they have a reason why you're wrong. "It's a condominium." "It's a rental." "We don't like Chicago."

A short list of banks that had expressed strong interest was getting shorter by the day, when out of the blue, I got a call from Rod Colburn at Goldman Sachs.

"We'd like to do the loan," Rod said.

"You want to take a piece of it," I clarified.

"No, lead it."

"Lead it?"

"Actually, we want to do the whole loan. We don't want any partners."

I was floored. I never imagined anyone would offer to finance the entire project, let alone an investment bank that was evolving into a regular bank. Goldman Sachs was the last place I would think to go to for a construction loan. But then again, the 151-year-old financial behemoth was trying to reinvent itself.

In the wake of the 2008 global financial collapse, Goldman, known for its high-octane sales and trading, remade itself from an unregulated investment bank into a commercial bank. It wasn't that they wanted to become a bank-holding company, but if they wanted government bailout money they had no choice.

The bank needed an injection of liquidity not only because it needed money but also to ensure market confidence. I remember at the time when the world's biggest banks were failing, a friend of mine at Goldman called for seemingly no reason. While we were chatting, he worked into the conversation that they weren't "nervous around here, because we carry $100 billion in cash at all times." When I got off the phone I couldn't help wonder why this friend had mentioned Goldman's liquidity to me. Although I never got confirmation, I imagined that in the wake of Lehman Brothers, Merrill Lynch, and Bear Stearns' collapse that Goldman's employees were encouraged to talk up their liquidity strength as another bulwark against a run on the banks.

After its reinvention, Goldman could receive money from the federal government to meet its liquidity needs and wound up taking $10 billion from the Troubled Asset Relief Program, the Treasury Department's program to provide capital to lenders in order to stabilize the economy. I did some business with Goldman after that, but the transactions were always somewhat difficult to close since it wasn't second nature to the bank.

So it seemed unbelievable when Rod said Goldman would finance *all* of the construction of 1000M. Although the biggest banks in America didn't want to lend more than $75 million to a single condo project, his bank was going to write a check for $340 million?

"I'm a little surprised by all this," I told him.

"We don't usually make construction loans," he admitted, "except for ultra-high–net-worth people . . ." Then he went on to give the names of three or four top developers and their projects that Goldman had financed. I had no idea.

"I'm going to send you a term sheet," he said affably.

Their terms turned out to be quite reasonable, good rates, deadlines, and so forth. The one unusual provision was they wanted me to personally guarantee 75 percent of the loan amount. Normally, a loan like this would require a 25 percent guarantee, but Goldman demanded higher guarantees on this and similar deals. Frankly, the whole discussion of personal guarantees was academic since I didn't have an extra 25 percent of $340 million in my checking account let alone $225 million. But they were the best game in town, so I said okay.

STUART BRUCK

Director of Mortgage Brokerage, Time Equities

My role for Francis is really to find financing for projects that he's involved in, whether it's buildings he's acquiring, buildings that he already owns that need to be refinanced, or buildings that he's going to construct. When we went out to the market with 1000M, it was already a point in time where there was a lot of other construction of upscale luxury condominiums in Chicago. There was doubt amongst the lenders, not as to our ability to put up the building, but as to the ability to sell these units. They thought there might be an oversupply of luxury condos in Chicago where they were not as in demand as they were in New York. It wasn't a commodity the way they are in

Manhattan where people spend outrageous sums of money on them. We did have a couple of quotes from a couple of lenders, and of course, we selected Goldman because they offered a superior rate. Francis has had a great reputation in the market. His integrity is beyond doubt, and he does what he says he's going to do; and that always counts a lot with lenders. Plus, Francis has been extremely successful over the years, and lenders are not afraid to lend to him because of his assets and his standings. The deal we ultimately struck with Goldman was for an attractive interest rate and terms. But Francis had to meet their guarantee requirement, which was higher than the industry standard at that time. In other words, other lenders at that time would charge a higher interest rate and other more difficult terms but have a lower level of a personal guarantee.

FRANCIS GREENBURGER: Another huge, mind-numbing part of the development process is the legal aspect. These are some of the densest, most difficult documents one can imagine, and if you don't read them with a microscope, you never know what's buried in there. We went through the documents for 1000M with a fine-toothed comb, because this was not a standard first mortgage or construction loan. Whenever something is written in what I call "high legalese"—defined terms that can take on a whole other elaborate meaning to render a paragraph, in my opinion, indecipherable—I insist the language be understandable. If not, I won't agree to it. I only do this in special circumstances, but 1000M was a special circumstance because the documents were full of twists and turns. The commitment letter for the loan, which describes the document to come, was a hundred pages long!

DAVID FEINBERG

General Counsel, Time Equities

In the Time Equities world, what we do here as in-house counsel is perform all of the legal work for the transactions, which can be anything from buying a property, selling a property, financing a property, and refinancing, leasing, operation agreements, and so on. Basically, it's a lot of paper, but without it, deals can't happen.

I started on 1000M with then-general counsel, Phil Brody, who unfortunately passed away in 2022. During the acquisition of the property, I worked on the operating agreements between the entities, the tenancy and common agreements, and also negotiated the partnership documents with our partners and outside counsels. We have to make sure all the closing documents are prepared properly. There are all these extra fees and costs at closing that you have to go through as well as specific provisions within the forms to make sure your client, which in my case is Time Equities and Francis, is legally protected in the deal.

Once we had the property, the vision had to be explained to lenders. There was a building to tear down and other preliminary development costs to eventually build a tall residential tower that needed funding. This was before we even tackled the construction loan. The banks—First Citizens Bank and then Goldman—wanted certain protections or restrictions on the property. If something should happen, their recourse is the property itself. So they want to make sure that you're actually proceeding with all the developments. We had to develop budgets and get building permits, making sure the bank approved everything we were doing.

One issue was that while we owned 1006 South Michigan, the building next door to 1000M, we had to secure the air rights over it to expand the tower once it cantilevered over 1006 for more floor area for the upper floors of the building. There was quite a series of exchanges over that contract. Helmut Jahn's design was not easy to explain to other lawyers.

Eventually, we went ahead and got a $300 million construction loan with a number of banks where every number on every document for the closing needed to be reviewed and re-reviewed.

Legally, a lot goes into constructing a building like 1000M. You have to protect the investors' rights, the partners' rights, the banks' rights, the city's rights. There are a lot—a lot of moving parts. And, as I said, *a lot* of paper.

FRANCIS GREENBURGER: You have to have a very substantial net worth—as well as experience—for the banks to do business with you on a development project of this scope. The original idea was for JK Equities and Time Equities to be fifty-fifty partners. Even though the Karliks had done big projects, by modern-banking standards he was still considered too minor a developer to guarantee a loan of this size. There was another issue that came up pretty quickly: They didn't have their share of the money to fund the building. The main sources of Jerry's capital for the project were regional investors, Elias Abubeker and Mindy Ouyang, who he partnered with on other projects in the area.

MINDY OUYANG

1000M Investor

I had been a mechanical engineer in China before I immigrated to the United States in 1991 at the age of thirty-eight. My husband, a computer science engineer, and I made the decision to come to America because we wanted a better future for our daughter. We didn't want our child to have our life.

We landed in New York where, for a year and a half, I attended language school part-time to learn English while working for a small architectural office in Queens run by another immigrant, who came from Greece. I drafted construction drawings, drawing on knowledge from my father—an architect and construction engineer—who was very famous in his hometown.

While working in the architect's office I noted the extremely long lead time necessary when ordering windows. Why did it take seven months to get a window? I was shocked—they are not hard to make. I saw an opportunity to put my engineering knowledge to use and started Monda, a small manufacturing business making doors and windows. Everyone thinks Monda is a play on my name, Mindy, but they are wrong. I was standing at my window of the small factory building we had rented in Corona, Queens on a Monday when a random thought popped into my head: "Everything starts on Monday." At that moment, I saw a Honda drive by and it hit me—the name for my business—a mash-up of Honda and Monday, Monda.

The United States was everything I hoped it would be when I was in China, a land of opportunity. But my husband didn't like it here; he wanted to go back. I wanted to stay. So we divorced. I wound up remarrying a man I met in my English language class, who was also a mechanical engineer by training.

Meanwhile my manufacturing business was doing very well, something one of my neighbors noticed. The older Italian man, who ran a general construction company out of the corner building, was a fan of mine. When he was ready to retire, his wife approached me about buying his property for a million dollars. I didn't have a million dollars! But yes, I was successful. In business for a little less than two years, I already had $500,000 in the bank.

"I'm only going to sell to you," my neighbor said. "It doesn't matter how much."

He explained to me, however, that I could borrow money by getting a mortgage. I didn't think about going to the bank to get money. I was an immigrant and didn't know. In the end, I spent $750,000 to buy a corner lot in Queens that is now probably worth $12 million.

There are all kinds of surprises in the United States—Chicago was one of them. I discovered the city while attending a trade show. The snow stopped me. I told my husband, "I want to go to Chicago because of that snow. The white color calms my mind." This was in the early 2000s when there was not a lot of manufacturing happening in Chicago. I opened a second window and door factory there.

In 1999, I met Eli, who was working for a large construction company overseeing a public housing project. After seeing our truck, he called me and interviewed me for the job of installing the windows for the development. He thought I was a hard worker, and my prices were good (he actually cut my price very low). Eli gave me my first public housing contract for over two thousand windows.

There were a lot of government construction projects to be had, but I couldn't handle the paperwork myself because of my English. I recognized that Eli was a gentleman who understood that world, so I invited him to join my company. Having seen the quality of my work as a project manager when he subcontracted the windows out to me, Eli agreed to become my partner.

At that time I had money. I had two factories with almost two hundred employees. I really had money, cash money, in the bank. Eli couldn't believe it. "Why do you have your money sitting in the bank?" he asked me. He wanted to become a developer because of the potential for big profits and encouraged me to get into the business. I met Jerry Karlik and invested in his development projects. Between 2006 and 2008, I lost money like everyone else. In my case, it was almost $10 million on real estate.

If you want to make big money you need to lose big money. Despite those losses, I like owning real estate. I'm from a certain generation where I don't want to be a billionaire of things I can't see. I don't know about the stock market, and I don't want to learn about it.

When Jerry approached me about investing in the conversion of a South Loop loft building into apartments on 2100 South Indiana

Mindy with Francis on a site visit

Avenue, he was working with another real estate firm: Time Equities. I studied Francis's company and saw that he is very successful. I also read his memoir, *Risk Game*. If Jerry hadn't brought in Francis, I wouldn't have gotten involved. The conversion into a mix of residential and commercial real estate, which opened in 2013, turned out to be a very good investment.

I respected Francis even more after I got to know him better through 1000M. He doesn't look down on anybody. Even though, when in comparison to him, I am a very small investor, he treats me like a partner. He comes to us for meetings instead of making us come to him. And at the meetings, he wants to hear everybody's voice. He is open to advice from all kinds of sources. That's how I know he is smart. Every time I talk to Francis I learn something new from him.

I have enough money to last the rest of my life; I don't need more. The reason I wanted to be a part of 1000M is because it keeps me learning something new about a world that I love. I live in the South Loop, and on the weekends I take water and something to eat and spend the day walking around. I love the buildings here in Chicago—there are so many stories.

FRANCIS GREENBURGER: We had already each put in about $12 million in equity by the time Goldman came into the picture, and they wanted me to guarantee 75 percent of the entire loan. I wasn't going to be principally responsible for this high guarantee and give the Karlik group half of the profits—unless they could produce an ultra-high–net-worth person to back their half of the guarantees. At one time, there were certain lenders who would give these construction loans for higher interest rates without guarantees. That wasn't part of my world, but I had heard about them. If Jerry, Mindy, and Eli could find a nonrecourse lender, I was all for it. But those kind of loans, when they were available, were for lower percentages of the project costs and had mostly evaporated long ago. So, we ended up where I knew we were headed. I agreed to sign my name, guaranteeing $255 million to Goldman in exchange for 75 percent equity ownership—plus a fee for guaranteeing the loan.

No sooner had we ironed out our partnership and decided to move forward with Goldman Sachs than the investment bank changed the terms of its offer. About four months after Rod first approached me, while we were in the final approval stage, the loan "committee" at Goldman decided it wanted us to find participants in the loan after all. They would only give us $200 million toward the total loan amount. It was frustrating but not surprising. When building a skyscraper, anything can and will change—at any time.

Goldman's share of the loan was still gigantic, but the news was nonetheless crushing. To get other lenders to buy in with Goldman for the rest of the $143 million posed a whole new set of challenges. Because Goldman's way of doing business is not a natural fit with traditional lenders, most banks might not want to partner with them on a big development project. We had been counting on receiving the full loan from Goldman, and then the game changed. We mastered the rules; they changed them—that *was* the game.

ROB SINGER: Goldman Sachs was unique in the market. Whereas other lenders did not want to make construction loans, Goldman did—and better yet, they wanted to make large construction loans. Our hope on 1000M was that Goldman would provide the entire $343 million construction loan that we needed. But as their approval process wore on, it became clear that they would lead the loan, only funding up to $200 million. We would have to find other lenders to participate in the financing to fund the remaining $143 million. The news was depressing. Finding lenders to participate in group loans is really difficult, and in this particular situation, it might have proved impossible.

Francis with Rob

FRANCIS GREENBURGER: When I had first sought out lenders for 1000M, I had asked M&T to participate. Based out of Buffalo, New York, it was only one of two banks in the S&P 500 index that did not lower its dividend during the 2007–2008 financial crisis—and it is one of my closest banks. M&T, with branches along the Eastern Seaboard, was at first reluctant to get involved because Chicago was not one of their lending areas and since they like to lend closer to where they have bank branches.

However, because of our long and profitable working relationship, they said if I had a problem with the financing to let them know. It didn't take me long to call and say, "I got a problem."

As I suspected, other banks were worried about Goldman. In large development projects that require multiple financing sources, the bank that puts up the most money is the leader. Other banks, however, didn't know what Goldman's skills were in this regard. I realized I needed a well-known construction lender used to syndicate their loans like M&T to step in. Because I'm a valuable customer, they agreed to take a hard look. The conclusion: They would put in $35 million.

The sum, while not large in context, was very meaningful to the market. In addition to the loan, M&T's syndication department, which regularly deals with other banks to participate in their loans, would help me find other banks to fill in what we needed. While my long track record with M&T was what got my foot in the door, the merits of the project ultimately sealed the deal. Their whole lending group took a trip out to Chicago, and once they saw the project they went from reluctant to extremely enthusiastic and supportive.

STUART BRUCK: We were dealing with the people in the personal banking division at Goldman, and they did not have a lot of experience in ground-up construction. Bringing in M&T to participate in 1000M was a move that other lenders seemed to favor, because now there was somebody in there who would be really competent at administering a construction loan. There's a lot involved in administering a construction loan because you draw down money from the bank as the work is completed. The lender in charge of the administration needs to send out engineers to check that 100 percent of the foundation is completed, plus 20 percent of the electrical work and 25 percent of the plumbing. Dealing with all different trades gets complicated quickly. There were six lenders when we did 50 West Street, and when issues came up in terms of project costs running over, all six had to agree, which was frustrating. It's like Jenga: If one lender pulls out, you hope the rest of the financing stack stands. You're just really hoping that all the pieces come together, and stay together.

Chapter 5

THE WISDOM OF THE MARKET, 2018

When Condo Sales Disappoint,
The Team Comes Up With a New Plan

ROB SINGER: "This isn't really working," I said to Francis at the end of 2018. A year after opening the sales gallery and spending more than $30 million to design and market the project, we had sold contracts on maybe seventy-five units at an average price that was at or below our cost.

Selling contracts to purchase condos on a preconstruction basis is often a difficult task, even in the best markets. This is especially hard when the building is not even under construction, which was the case with 1000M. You're asking people to commit to a major home purchase years before they will be able to take delivery.

The preconstruction contract purchaser is generally motivated to take the risk if they believe that they're getting a great deal. They want to cheaply purchase a unit contract and then hope that the value increases substantially by the time the unit is built and delivered. The problem at 1000M, among other things, was that the market did not view our average preconstruction prices as a "deal."

Micro-unit rendering

We were all worried. In the hopes of generating some sales momentum, we introduced a new product: the "micro-unit." We significantly redesigned the unit layouts on about ten floors of the building to be much smaller and more efficient. Instead of eight units on a floor, we had sixteen to eighteen units.

The idea was to shrink the size of the units so that we could also reduce the gross sales prices (although this would increase the sale price on a dollar-per-foot basis). We thought buyers might be more willing to sign preconstruction contracts on a 850-square-foot two bedroom for $850,000 (or $1,000 per square foot), even if they didn't seem to want to purchase one of our standard 1,400-square-foot two-bedroom units for $1,260,000 ($900 per square foot). And maybe if these units were a hit, then they would create sales momentum for the rest of the building.

"Efficient" units like these were common in Asia and Europe, so we directed sales and marketing efforts to these markets as well as within the United States. But it didn't really work—we sold contracts

Micro-unit rendering

on a few, but they certainly were not a hit, and they certainly didn't spur sales of regular units in the building.

The sales and marketing agent for 1000M, @properties, did not agree with our decision to launch the micro-units in the middle of the sales campaign. @properties argued that micro-units would signal weakness to the market, create an additional load on the amenities and elevators, and frustrate the few contract purchases they had been able to attract.

In the end, @properties was correct. Some of our larger contract purchasers cancelled their contracts upon hearing about the micro-units. I don't think we regret trying this program—we had to turn over every stone.

JAVIER LATTANZIO

Director of Sales, Time Equities Brokerage

A year after 1000M started sales, they hit a wall and for some reason, couldn't sell anymore. Francis asked me if I could assess what was going on because of my experience running the sales for 50 West. We opened sales of Francis's first skyscraper in 2014, two years before it was completed, and were met immediately with insane success. For example, on the very first day we sold three out of ten penthouses, with another three reserved. We continued to sell at a ridiculous rate even as we increased prices.

While 1000M was in the opposite situation, it shared one major thing in common with 50 West: We were selling paper. If the site is just a hole in the ground, my job as a salesperson is to make you fall in love with your floor plan. The first step is to figure out the buyer's needs—one bedroom or two? Water views or lower level? You can't start by showing a potential buyer seventy apartments available for sale. Most clients in a sales gallery have very little attention. If you don't narrow down the options, it becomes very difficult for clients to make a decision.

You keep repeating—in different ways—the question "What price point are you in?" until they finally tell you. Once they do, you show them the type of apartment that's in their range, and more often than not, they want something a little better and are willing to go a little higher.

"For $3 million, this is what you can have."

"What if I can pay four?"

"OK, $4 million; this is what you get."

By the end of the conversation, you've landed on an answer. They want the $5 million condo overlooking the water. Then you show them *two* options, not twenty. "Maybe I have more, but they are not released," I say.

Micro-unit with separate bedroom area

"Do you have something higher?" The client suddenly rethinks.

"I may have something higher."

"How much higher?"

"Ten floors higher, but it's going to be also $2 million more."

"$7 million to buy the fifty-seventh floor? Yes, I can do that."

That's an example of how it happens. Then we put the client in a happy room with a drink and talk about which floor plan they decided to go with. That's where you have to close. The industry average is that it takes seventy people to come into one site to make a deal.

In winter 2018, I went to Chicago to assess what was going on. I did my market research, including going to other sites, and in the end thought that 1000M was a good product. It wasn't a home run since the prices they were trying to achieve were high for the area. But I felt the numbers were workable—if there was more pressure in the sales office. In my opinion, they weren't pushing hard enough.

I thought the team was good. They knew what they were doing and how to sell—they just needed better management. For example, I was shocked to find out that the sales office was closed on Mondays. When I asked why, they said it was because they worked Sundays. With eight people, there was no reason to close the office. They could rotate schedules, especially if they worked as a team.

Even though before becoming a real estate broker I competed professionally on the international tennis circuit, I like to be part of a team. Although there were only three of us selling units at 50 West, I convinced Francis to let us work as a team as opposed to individually. Instead of a model where the agent who makes the sale earns the commission, we all made a base salary and divided up bonuses for each sale we made. The result was that we didn't compete internally or worry about losing a sale if you took a day off. The whole operation ran more smoothly and efficiently because we helped one another. I wanted to bring the same model to 1000M.

"The only way this is going to work out," I told Francis, "is if I go there myself every week to start asking questions and putting some pressure on them."

Shortly after the holidays, I began traveling to Chicago once a week to work with the 1000M's sales team. Here I was, this fucking guy coming from New York to push people to do something that they already knew how to do. I'll admit I'm a pain in the ass. I'm not the easiest person to work with because I want to make sure that we're doing the best we can to get the results. From week to week, I followed up on the progress of every deal, analyzing the ones that fell through, troubleshooting the ones in jeopardy, and celebrating the wins with incentives like an all-expense trip to New York.

I wasn't just putting pressure on the sales force and buyers to come up on their offers; I was also pressuring the developers and investors to agree to lower offers to less than they initially wanted. If a market is very hot, you'll get the asking price—or higher. But in general, people want to negotiate, especially in the kind of market we were in. My job is to understand where everyone can meet so we can make the deal happen. I'm a negotiator—that's really what I am. My job is to find a happy ending for everybody.

FRANCIS GREENBURGER: More than eight months after introducing the International Collection, the sales were still lukewarm. While the number of contracts had risen to nearly one hundred, most of them were at relatively low prices—low enough to make the financing unattractive if not unfeasible. Goldman Sachs was continuing to give mixed messages about their construction loan. Some days it seemed like they were moving ahead; other days, they weren't so sure. Time was not on our side. In the sales contracts we had stipulated that we needed to start construction by December 2019. Otherwise, everyone could get their money back.

Meanwhile, 1000M's October 2019 groundbreaking was right around the corner. Nervous about the largely ceremonial event scheduled for October 24, I met with Rob to voice my fears. "I'm not sure we're going to know where the financing stands within thirty days," I said.

If the puzzle of financing a development on this scale didn't come together by December, then it was official: I had a problem. I wanted to know how much money we had already sunk into the project if I needed to pull the plug. As Rob explained, our biggest expense coming up was not assembling the equipment needed for real construction but, believe it or not, insurance. We had to prepay insurance for the whole job, which in this case was upward of $7 million.

Our insurance team met with the insurance people to see if I could take out a policy for construction of the foundation alone, then pay the rest before we went vertical. While they were fussing over my request, we made the decision to move forward with the groundbreaking. In the face of great uncertainty, we planned to be all smiles in Chicago on October 24, because, as Rob says, "A developer always has to show strength."

Despite our best acting efforts, the *Chicago Tribune* sniffed out 1000M's precarious future in an article published on October 16. The paper was accurate when it reported that the upcoming groundbreaking of Helmut Jahn's tallest building in Chicago, and the city's largest condo proposal since the 2008 recession had destroyed real estate values, "does not guarantee that 1000M will be built."

Citing the example of Chicago Spire—a "twisting residential tower that would have risen to two thousand feet" but wound up "a hole in the ground just west of Lake Shore Drive"' when construction stopped after the project broke ground—the *Tribune* offered context that didn't portend well for our building: "There has been far more construction of rental apartments than for-sale condos."

"1000M's developers declined to disclose details about financing or the number of presold units," the article reported ominously.

Our PR and communications team went on overdrive ahead of the October 24 groundbreaking. In the *Tribune*'s coverage that day pegged to the event, we tried to turn the negative narrative around.

"Chicago's biggest condominium project since the recession will be under construction by December, and the Helmut Jahn–designed skyscraper will open within three years," the paper reported. "The $470 million project is within a month of finalizing the first of two rounds of construction financing led by Goldman Sachs, according to the developers."

In response to questions about starting such a big development so late in the construction cycle and despite warnings of a looming economic recession, I tried to project optimism.

"Building something like this is not something you wake up one day and do," I was quoted in the piece. "It takes a long, long time, and you have to be prepared to go through different cycles. Who knows what's going to happen with the economy? We all know it's flashing red lights. But our business plan allows us to weather whatever the environment is, because we look at this as a long-term project."

There was only so much I could spin. Facts were facts and, as the newspaper reported, 1000M was beginning construction with only 23 percent of the condos in contract. And out of those ninety-seven units in contract, thirty-eight of them were micro-units. To have racked up sales nearly $100 million in sales wasn't *nothing*, but we were far short of the amount needed to convince banks to give us a construction loan for 1000M.

Out of a total budget of $465 million, we as the sponsor wanted to put in about 10 percent, or $46 million (although the percentage would eventually rise to 12.5 percent). The other $420 million needed to come from investors and lenders. You can't use buyer deposits to fund construction because in order to protect buyers, the law requires deposits be escrowed until the building is completed and the purchaser closes on their apartment. Plus, to incentivize buyers to take the risk of purchasing condos early on in the process, we offered very low deposit requirements. Instead of a 10 or 15 percent initial deposit, we offered deals with only 5 percent down on signing and another 5 percent once we started construction. So even if we could have used buyers cash to fund the tower, there wasn't much of it. We had to raise the money by convincing capital markets that our building was a good investment by presenting the costs, sales, and projected profits.

That wasn't only true for the construction loan. While my team pushed that boulder up the mountain, there was a whole other uphill financing battle we had to wage to make 1000M a go.

For a development project, a typical capital stack—which represents the different sources of funding that will ultimately equal the total amount to complete the project—involves a first mortgage, second mortgage (sometimes called a mezzanine loan), equity, and sometimes preferred equity. A mezzanine loan is like a second mortgage, but we don't call it a second mortgage because that's a dirty word in the banking world. While banks typically provide the construction loan portion of the capital stack, hedge funds often offer the preferred equity at a hefty interest rate and complex terms. If our construction loan had a rate of around 4 percent, the preferred equity could cost as much 14 percent. With my 12.5 percent sponsor equity, the "mez" loan covering anywhere from 72.5 to 87.5 percent of the project would provide $90 million of the capital stack.

We only got nibbles from the preferred equity market, none of which materialized into anything close to real money. As October turned into November, the pressure mounted because of the crucial deadline only a month away. December 31, 2019—two years after we started sales—was when condo buyers could ask for their deposits back if they were getting cold feet.

The flashing red light was serving to make an already difficult process even harder. Necessity, as they say, is the mother of invention. In late November, with the construction-start deadline just around the corner, I got an unconventional idea on how to raise the preferred equity.

It all began when I first saw the rates the institutional preferred equity lenders wanted. They were so high that I said to myself, "This is crazy." Around the same time, a broker from one of these lenders called me to ask if I wanted to invest money with them. Elliott Investment Management, one of the world's largest hedge funds, who we had worked with on another project, was offering to make preferred equity investments on my behalf.

It struck me: Preferred equity lenders were getting money from people like me, relending it, and marking up the cost. Time Equities already had over two thousand individuals, who made traditional real estate equity investments with us. While we offered our investors an equity fund that represented our company's diverse acquisition strategy, I had an idea for a different product. It was debt and not equity—and it also had a clear exit in five years. If we went directly to them as the hedge fund broker had done with me, would they be interested in this type of investment as well?

First, though, I had to convince the managing director of our equity division, David Becker, that my unorthodox plan for financing the preferred equity could work, since it would be his job to put it into effect.

I thought he would jump at the idea, because David always likes to do more business, but he was uncertain. Although he has delivered for me many times over our decades of working together, David is not a yes type. When he agrees to do something, it was very important that he is able to complete it. He didn't want to promise me a result and not deliver. This was unchartered territory. Even though I knew he was willing to do everything he could to succeed, there was no way he could be certain about the outcome.

The other hurdle was the large amount of money that we had to raise in a short period of time. By way of comparison, we had just raised $130 million for one of our equity real estate funds from the same pool of investors over a two-year period. To raise the $90 million in nine months that we needed for 1000M, for a new construction project, was, to say the least, a challenge.

But I kept pushing David. I wanted him fully committed, because he is very capable when he wants to execute. Then I insisted. When we finished the presentation materials and he saw the whole project come together, David was on board, and right after Thanksgiving he went out to the marketplace to sell our skyscraper to others.

DAVID BECKER

Managing Director, Equity Division, Time Equities

Francis told me it'd be a mini miracle if I raised the $90 million this way but that he needed me to deliver. I know he has confidence in my skills, but this was going to be very hard to sell since it had never been done before. I had to convince sophisticated financial advisors, broker dealer officers, and investors that it made sense, even if it was totally out of the box. It helped that at Time Equities we typically go against the grain, thinking independently and presenting contrarian products.

Fifteen years ago, when we embarked on building a platform of investors, Francis said to me, "David, I don't want to create something that's not fit for who we are and what we do. Institutional partners are offering to fund us if we invest in the type of buildings and locations they are focused on. I want to sell Time Equities' investment strategies. We need to be who we are and offer others what has made us successful over long periods of time."

That's the essence of this company, which goes back to Francis, who built a real estate empire from nothing by using his courage and ability to see and create value where no one else could.

In building a group of individual investors and financial advisors over the last fifteen years, we wanted to raise capital from regular people as opposed to large institutions. The cost and flexibility of that capital allows us to execute a project and realize our vision and value for both Time Equities and the investors who partner with us.

We currently manage hundreds of financial advisors and thousands of individuals throughout the United States and abroad; our reach has enabled Time Equities to grow exponentially. 1000M, however, was the first time where we were offering a large-scale, high-rise development project directly to our private family office partners, investors, and financial advisors. We typically would offer more predictable type investment opportunities that had cash flow and some upside through various value-added features but were mostly stable.

This was a very big undertaking; but we gained some momentum, got our offering and marketing materials in place, and the plan started to come together. We had designed a very creative, groundbreaking structure with an approach to primarily target large 1031 tax exchange investors.

This was definitely a new milestone for Time Equities. If we could pull it off, this would be huge for the company. Not only would it mean millions and millions of cost savings on the cost of the equity capital but it would also be a whole new business line for the company. We could begin to incorporate some of our larger, higher-profile development projects into the investment opportunities we offered our investors.

We mobilized in December 2020 and hit the market in January. My initial view was that if we could gather momentum in early 2021 and close out the offering by summer, that would be a huge win.

I knew I was one of the last resorts for the equity that would fuel the project because the institutions were far too expensive. We built a phenomenal platform with very deep and reliable relationships that would benefit us, but the question remained whether or not I could actually make it happen. I'm a competitive person, and my team follows my leadership. I'm going to barrel to get it done. I don't want to take it on halfway. I think Francis, in his gut, knew I could do it. Still, it was going to be an uphill battle.

FRANCIS GREENBURGER: December's arrival meant we had a big decision to make: Do we start construction on a half-billion-dollar project with a hundred or so crappy contracts, half a term sheet, and no construction or mez loan? Or should we cut our losses, even if they were in the millions.

Anyone who knows me knows I don't like to lose money. "Let's start construction," I told Rob, "because I don't know what else we're going to do." I hoped that a demonstration of strength, in the form of $20 million dollars of foundation and utility work, would bring a flood of contracts—which in turn would bring the bank money.

"We're just going to will this thing?" Rob said.

"Welcome to the roller coaster of skyscraper construction and development."

Chapter 6

STOPPING IS HARD, SPRING 2020

COVID-19 Derails Construction

FRANCIS GREENBURGER: Less than two weeks after the World Health Organization declared COVID-19 a pandemic and the New York City public school system—the largest in the country—shut down, there was hardly an aspect of American life untouched by the new and deadly virus. 1000M was no exception.

At the end of March 2020, I was worried about COVID-19's possible effect on the health of construction workers who had been working on the site since late December. The official guidelines on how to minimize the spread of this frightening disease were completely unclear. The state of Illinois allowed construction to continue under the rubric of "essential work," but the City of Chicago had its own set of rules—as did the unions, who were understandably skeptical of the government's ability to keep their members safe. The nature of the construction at that particular moment also made it extremely hard to keep people six feet apart from one another. The next step was laying down the massive concrete-base slab of the building, requiring many people working in close contact to ensure the continuous and complicated process of a smooth and even pour. Up until this point, we had been spending about

$5 million a month on building the foundation, finishing the plans, and filing permits. To begin the pour we would need to bring in a vertical crane, a piece of equipment with a very expensive weekly rate. If we were to rent the crane, and the next week the governor or mayor shut down construction, we would be stuck paying for a pricy, unusable piece of equipment.

That brought me to my other major concern: money. The pandemic's effect on the financing process was as swift and sweeping as COVID-19 itself. Not surprisingly, banks were canceling their commitments right and left. At my real estate firm, Time Equities, we had banks pull out on half a dozen loans already in process. Even without a global virus threatening to upend the world order, managing the bank financing of 1000M was already—to put it mildly—difficult.

When COVID-19 hit, Goldman Sachs maintained its position that it planned to finance $200 million of the construction project, $140 million less than the bank's initial offer of full financing. M&T—a bank I had a long-standing relationship with—agreed to finance $35 million and help us syndicate with other banks to raise the outstanding balance. We were in the process of reviewing and negotiating with M&T the details of the documents needed for other banks to participate when the coronavirus brought everything to a screeching halt.

During the first two weeks of the pandemic, I didn't want to contact M&T in any way while they were dealing with a million issues. To process thousands of federal Paycheck Protection Program loans, the bank took every single loan officer on staff and reassigned them to work on these governmental PPP loans. A luxury real estate development in Chicago, I imagined, was the last thing on their minds.

I had been right. When I eventually communicated with M&T, they conveyed the fact that they weren't in a position to deal with our construction loan internally, let alone go to outside banks. Even if they tried to syndicate the loan for 1000M in that moment, they said they wouldn't be successful since nobody was looking for new commitments. I agreed with the news, which was not all doom and gloom. In terms of M&T's participation, the bank's board signaled that they would most likely go through with the $35 million loan when they could return to regular business.

Even if M&T put up $35 million and Goldman Sachs came through with $200 million—which it conveyed was its intention—we were still short about $100 million. While we weren't going to get that from other banks anytime soon, we had started construction and we needed to close a construction loan.

This was the backdrop to our next proposition to Goldman: lend us the missing amount at a higher interest rate. We were asking them to become a "standby lender," which charges a high interest rate to issue a loan they initially didn't want to make. Many construction loans don't get syndicated before they're entered into. As an example, M&T loaned us the full $75 million to build a project in West Palm Beach but later, after the loan closed, partnered with other banks.

Even if the world was falling apart around us, we needed the money and commitment right now. We needed Goldman to step up and close the entire loan because no one else was going to come to the rescue in the midst of a pandemic. By 2021, we presumed the world would be in a better place. At that point, we would help Goldman sell down the loan by finding partner banks.

My team set to work on presenting financial information Goldman requested to support our proposal. It showed that despite the chaos of the pandemic our fundamentals were still strong. If the numbers weren't proof enough, there was the matter of the investment banking firm's reputation. Goldman Sachs had its corporate logo on a sign on our site. The media had reported they were financing the project. If all of a sudden the project literally collapsed because Goldman wasn't able to finance it, we weren't the only ones who were going to look dumb.

These were extraordinary circumstances, and nothing was normal. The unknown variables were so many that I was conflicted about convincing Goldman to give us the construction loan for 1000M. In a funny way, if the investment bank turned us down, the decision would be made for me. It would be painful to lose the money I had already invested, but we would have no alternative other than to shut down the site. However, if Goldman agreed to move forward, the onus fell on me to find the rest of the money in this unprecedented climate.

That included the mez loan or preferred equity. I had decided to do something unconventional for this part of the capital stack provided by hedge funds, or big financial funds at very unfavorable terms. Rather than looking for one big investor for the $70 million mez loan, I decided to see whether some of our midsize investors would like to invest anywhere from $500,000 to $1 million or more as part of a syndicate. We had raised $25 million when COVID-19 brought everything to a screeching halt. Not bad for an unfamiliar approach, but we still needed another $45 million.

Last but not least, I personally had to come up with roughly $60 million for the developer's equity portion of 1000M. While I had already put in about $40 million, I was, to say the least, uneasy about sinking another $20 million or so of my money into the project while the world around me fell apart. Hence my ambivalence about pushing Goldman to take a gamble and close on the entire construction loan: Because then, the pressure would be on me to come up with the rest, and I honestly didn't know if that was possible in light of current events.

ROB SINGER: With respect to securing a construction loan, our easiest path forward would have been for Goldman Sachs to fund the entire loan, instead of requiring other lender partners. Finding lender partners was extremely difficult, nearly impossible, given the short time frames that we were working with, not to mention the complexity of the deal.

We had sold contracts on maybe one hundred units. We had the permits and the plans. But we had no construction loan. Still, we decided to proceed with foundation construction. This was extremely risky, but it was also extremely risky *not* to proceed with construction, because that would have meant giving up on the project. We kept hoping that by demonstrating strength and progress, we would create our momentum, which would ultimately lead us to success. At this point, though, we really needed the loan—and it was not looking good.

FRANCIS GREENBURGER: On June 3, *Crain's* Dennis Rodkin, who reports on the Chicago-area real estate market, left a voicemail with our PR firm that set off alarm bells.

In April, we decided to shut down construction for a thirty-day period. Upon advice from our construction managers, it was our own opinion that the best thing to do was to pause while the contractors worked out a plan for COVID-19 safety protocols, such as masking and temperature checks.

It sounds easy, but stopping construction is really hard. Especially on a project of the magnitude of 1000M, unwinding work already in progress is a monumental challenge. The communication alone—calling all the various contractors to get them off the site—is a feat in itself. Understandably, there was a lot of confusion on what the contractors were supposed to do—when and why. Those weren't easy questions for us to answer. The only response we could offer was that because the workers' safety was of paramount importance, we weren't comfortable moving ahead.

That was absolutely true, but equally true was the fact that we were at a critical moment in the project. Beginning the concrete pour, an extremely complicated process, would activate the start date of the construction loan—which we still didn't have secured. Not only was Goldman Sachs still mulling over their participation in 1000M, they continued to increase the level of my commitment. Although the banking arm of Goldman is kept separate from the investment side, the firm's philosophy is that it should use its banking capabilities to serve its customers, who also use Goldman's investment services. With that understanding, at the start of the process we had created a small investment account with Goldman, which we beefed up with some one-off transactions to meet their standards. All of a sudden, a week before the call from *Crain's*, Rod, our contact at Goldman, reported that some people at the firm raised the amount of my investment business activity with Goldman as an issue.

The request felt like a stall tactic, which I fully understood since we were engaged in our own strategic delays. As I said, we halted construction because we didn't want to see construction workers infected with this new and lethal virus. But we also happened to be out over our skis in terms of securing the loan to move forward. Dennis Rodkin suspected an ulterior motive when he learned that construction on 1000M had halted.

Rodkin—a seasoned journalist with a decades-long history of reporting on the local housing market—was the one who had caused a big drama around the time of our ribbon cutting when he wrote a piece questioning the viability of our financing. Now he was asking the same question.

As soon as our PR person received his voicemail inquiring about construction at 1000M and his plans to write a story on its current status, she reached out to the team to suggest we provide him with a written statement so he had the facts. But what exactly were the "facts"? It seemed like every minute some new and dramatic element was added to the story. Some questioned whether we should comment at all. Despite the lack of intel on Rodkin's plans for a story, our PR group wanted to make sure he had the "right messaging." And because *Crain's* weekly real estate section was only a few days away, time was of the essence.

The easy answer to give the press was COVID-19. I wanted to say we would restart once the banks and we felt the situation was fully stabilized. But the sales and marketing folks were adamant the statement did not include any mention of financial institutions whatsoever. Their goal with this statement was for the reporter to decide there was no story here. They worried that hinting at any kind of problem with the project invited more investigation. Meanwhile the world was falling apart all around us. As someone who likes to be very up-front, I asked, "How long does the COVID-19 story last?"

In the end, I agreed to leave out the word *banks* in the statement that explained the concerns over COVID-19 but also highlighted the fact that not all work had stopped on the tower.

> 1000M completed the first phase of the foundation work. When the COVID-19 situation arose, we had workers underground, working shoulder to shoulder, making it effectively impossible to maintain proper social distancing. We consulted with our experts and we all felt the safest route was to temporarily halt construction on site to ensure worker safety. That said, we have and continue to advance all off-site critical path work, as we wait to remobilize on-site construction. These off-site activities include signing sub-contracts, procuring and testing the curtain wall, processing our building permits and finalizing construction documents, among other items—which allow the project to generally stay on track. We will resume work on site when we feel the situation has fully stabilized.

In the context of the statement, it would be understandable to think that the "situation" in the last sentence was referring to COVID-19. In my meaning, however, the situation included everything from the social unrest growing on the hot summer streets of Chicago to the reluctance of banks to lend money.

As it turned out, the "situation" didn't look like it was going to stabilize anytime soon. Only a few days after we released the statement, we heard back from Goldman: They were not prepared to go ahead with any part of the construction loan for 1000M until they had a better feel about the world at large. They advised us to get back to them in a couple of months when hopefully order was, if not fully restored, at least heading in that direction.

I had been handicapping Goldman coming through with the loan at 50-50, so I was hardly surprised. From a banking point of view, the world was too uncertain. Frankly, I shared their concerns. Though their decision to pull back messed up the project in many ways, even I had to admit that a hiatus probably made sense.

The immediate task at hand was disseminating the information about the delay in construction to the stakeholders directly affected by the news. They included not just the many contractors who had to put a hold on their work but also the buyers. Their contracts stipulated that until we blew the delivery dates for their apartments starting in 2022, theoretically they couldn't get their money back. However, it was a goodwill gesture to avoid lots of angry people.

When the condo buyers originally signed their contracts, they put up a deposit of 5 percent. When we began construction, the contracts stipulated they had to put up another 5 percent. Maybe I would return the 5 percent that they put up for the construction period with the understanding that when we actually started construction again, we would ask them to repay it.

In a way, this was the least of my problems. Whatever the terms we decided upon, the news would imply to buyers delays in constructing 1000M, and general uncertainty—a state that could be ascribed to every aspect of life during COVID-19. It wasn't only the pandemic that was upending society: The murder of George Floyd by a Minneapolis police officer on May 25 ignited protests and civil unrest throughout the country. Chicago was no exception. Five days after Floyd's death, downtown

A protest in Chicago on May 30, 2020

protests turned into violent looting that lasted for three days and spread across the city. Chicago became one of a dozen major American cities to declare a curfew, which virtually shuttered the downtown area. On May 31, Mayor Lori Lightfoot asked Illinois Governor JB Pritzker to deploy the National Guard to Chicago for the first time in more than a half century.

I was fully prepared that some 1000M apartment owners might want all their money back. We could cancel contracts and return the money. Those decisions would most likely have to be made on a one-by-one basis. Scrambling to deal with this major setback, we were prepared to do whatever we could. Although just what that might be was anyone's guess. It was painful, but it was a painful world.

While I might not have been shocked by Goldman's decision, I did find it depressing and as a symbol of the chaos threatening much more than just a tower overlooking Lake Michigan. I wasn't excited for the meeting I called the following Monday to hash out the plan for notifying the architectural team, construction management, sales team, and buyers. That meeting was followed by a planning meeting for Time Equities to look at its existing portfolio, as well as all kinds of new opportunities.

Instead of mourning this new COVID-19 era, we discussed how to move forward in it.

One of those new investments was in an office complex in Maryland. Located near defense bases, the tenants were mostly defense contractors. The building, which had lost money because of cuts to the defense budget, originally had a mortgage with M&T for $28 million. In financial trouble, the owner agreed to sell us the property for $16 million in cash a year earlier. The plan was always to put a mortgage on the complex, but I wasn't sure if we would be able to get one on an office building when many people were no longer going into the office. M&T gave us a $10 million mortgage on the property. It was good story for us: a bank loan completely done during COVID-19 proved there were parts of the world still functioning.

Rob—who fielded the initial call from Goldman about its decision—was already talking next steps. He was going to look at an alternative, a 100 percent rental scenario for 1000M. But that meant we would have to reduce the budget of the building substantially in order to make enough of a return. A 6 percent return once completed and leased up is the minimum threshold one would expect when building a new rental, although ideally it would earn 7 or 7.5 percent. Rob had a big project ahead of him, looking at current rents and projections, to figure out the numbers. It wasn't only about crunching numbers; there was a whole rethinking about the building he would have to do.

ROB SINGER: The financial markets were not excited about making loans or investing equity in 1000M. We kept arguing that once the project was financed and construction was underway, the "flood gates" would open. No one was particularly impressed with our business plan or contract sales to date, which forced us to start thinking hard about changing or abandoning it altogether.

We had the same situation with 50 West in Manhattan, where we had a business plan to build a hotel at the base of the building and small condos on top. Then, right as were starting the foundation, the Great Recession hit. We had to stop everything and come up with a new business plan. It took five years before we could restart. We salvaged the exterior geometry and architecture and redesigned the interior of the tower to remove the hotel and create larger condo layouts.

After substantial time and energy—not to mention massive risk and investment—we finally landed on an economic design and business. This allowed to secure a construction lender group that loaned us $300 million. The problem then was that we still needed an additional $120 million of cash equity before the lender would start to fund the project.

We scorched the earth trying to find this equity investor. Equity investors did not seem to believe in the project, or us, or both. TEI had never done a project like this before. We were not only coming out of the Great Recession but the development site was located in the Financial District, which had just been swamped by Hurricane Sandy. Everything seemed wrong. We ended up hiring a capital broker who identified an investor, Elliott Management. The huge hedge fund based in Midtown wanted to do the deal and weren't bothered that no one else wanted to do it. The money they were lending to us came at a high interest rate. Furthermore, the deal specified that they had to get all of their money back—and all of their interest—before we could get a dollar. It was the only option we had, so we took it. They ended up being great partners—very smart and flexible. They also made a significant profit, so it all worked out.

Francis has a million similar stories that reinforce the idea that when the world doesn't want to lend to you, that's a sign that you may actually be on the right track. Lenders and capital providers are often on the wrong side of the curve—meaning that they lend or invest at the end of economic cycles, when they shouldn't; and they don't lend or invest at the beginning of cycles, which is when they *should*.

In general, our experience with large projects is that they're not linear. They go through various ups and downs, market cycles and designs, before they finally click. 50 West went through ten years of redesigns and market cycles before finally launching and succeeding. It takes extraordinary financial and mental perseverance.

When we're working on these deals there is a lot of pressure. We're simultaneously designing the project, advancing the construction plans, securing permits, sourcing subcontractors and materials, searching for capital, and so forth. An enormous amount of money and risk are on the line. Oddly, I don't think any of this keeps Francis up at night. I don't know why. It keeps me up, and it's not even my money.

Toward the end of 2019, we were trying to project confidence and progress with 1000M, but we knew it wasn't going well. Obviously, it's disheartening when you've invested so much time and money into a situation and it just doesn't work. But I couldn't sit there and cry about it. You must keep working, keep thinking of creative solutions and next steps. There is no "off-ramp" once you're so far in. There's no telling Francis, "I don't have any more ideas. I give up." That's not an option.

We had $40 million ($25 million in expenses and a $15 million mortgage) invested in the property—with little to show for it. We had to make the project happen. I started to think about abandoning the condo business plan and switching the project to a rental building. Rental projects are generally easier to finance, although nothing in this business is easy.

FRANCIS GREENBURGER: Not everyone was convinced that Chicago condo sales were a no-win situation. Long after his involvement with the sales of 1000M was over, Javier Lattanzio, Time Equities' director of sales, still believed it could have been a success—if only people had tried harder.

JAVIER LATTANZIO: As much as I wanted to create a team in Chicago, at the end of the day, I couldn't convince the sales team that I was on their side. Every time I made a suggestion, the head of the team opposed it. He was unnecessarily threatened; I was never planning to move to Chicago—that was not my goal. This project could have been very successful; I still believe that.

In the first week of December 2019, I took the sales team out to a steakhouse for our holiday dinner. "Thank you so much for a great year," I said. "I think it was fantastic." And it was. Over the ten months that I had been traveling to Chicago regularly, we had made fifty-four sales. That was a number to be proud of in a very tough market. Still, I thought we could do better. "I want to make a change," I told the group around the table. "I'm not changing you, but I want to diversify the team a little bit."

The sales office was comprised of seven female agents and one male manager. If possible, it always makes sense to have a mix because not everybody responds to men and women the same way. Maybe it wasn't the right thing to say, or I shouldn't have told them my thoughts. Although no one said anything that night, the manager and the owner of @properties later called Rob to complain that every time I went to Chicago, I was disrupting the team.

"Maybe you shouldn't go there anymore," Rob said.

"Fine, I won't go anymore."

We brokers are the face of the company and that's why, for me, I can't accept mediocrity. I can't take it. It drives me insane—*insane*. You should be on the top of your game, no matter the market conditions, no matter the price point. My mission when I went to 1000M was to sell out the project, not for me but because that was the job. Together, we made fifty-four sales. The number of apartments they sold from December 2019 to March 2020? Zero.

Chapter 7

A RENTER'S MARKET, FALL 2020

Turning Condos Into Rentals

ROB SINGER: It was clear that nobody in the market wanted to purchase high-end condos at 1000M, at least on a pre-construction basis. But maybe people would pay sufficiently high rents to make a rental building economic? Anecdotal evidence was beginning to appear that some renters in the market would pay high rents. People seemed to be paying the most money to rent condominiums directly from owners. Several projects in a few cities delivering large rentals with high-end finishes and appliances typical of a condo demonstrated that people were willing to pay the high rents for the privilege of flexibility. We also saw this trend at 50 West, where we had two sponsor-owned penthouses, which I never thought we could rent out. One of them was ten thousand square feet, which at $100 per foot came to $1 million in rent per year, or $83,000 a month. We settled on listing it for $70,000 and wound up settling for $65,000 with a renter. Shortly after that, we rented the other penthouse, which is seven thousand square feet, for $40,000. Those were extreme examples, but the condo-style rental market was clearly real. The question, as we contemplated taking 1000M in a very different direction, was the depth of the market, which was unknown, particularly in a place like Chicago.

FRANCIS GREENBURGER: The first thing that happens in a building's life is the construction loan—what Goldman now didn't want to give us. That's how you build. The fundamental question for anyone giving a construction loan is how are they going to recoup their money when the building is finished. If it's a condo building, apartment sales pay down the loan. Once a rental building is finished and leased up, there are many banks that would want to make a loan on it. Rental real estate is of the most liquid financing markets around.

At 50 West, were able to finance a combination of rental and condo apartments. In this scenario, the financing for the rented units in a condo building is more expensive and less available than for a pure rental building, because lenders take the position that if you only own 50 percent of the building, you don't get to make all of the decisions, and that could be a problem. The way I look at it is different than the bankers, but the market is less familiar with that strategy than plain vanilla condo sales or full rentals.

Part of Rob's push for a rental came from an experience we had with another property where we learned there is no point fighting city hall or the market. At the same time we that began considering a rental model for 1000M, we unveiled the final plans for our big project in West Palm Beach, which had also gone through multiple redesigns before we were able to get financing. CasaMara—a resort community on ten acres that is two miles south of downtown West Palm Beach—had started out in 2015 as a fifteen-story, steel-and-glass condo tower on the site of a rundown commercial complex from the 1950s with a Sears and grocery store. The local government okayed the plan, but the neighbors came out with pitchforks. They didn't want tall buildings in their community. Although we had spent a million dollars on plans and countless hours in meetings and obtaining permits, we were shot down and had to start all over again.

By February 2020, CasaMara, on its way to construction, had morphed into three hundred units among seven low-rise buildings, as well as sixteen thousand square feet of retail space and a sixteen thousand square foot clubhouse with communal amenities such as a pool, co-working lounge, fitness center, dog park, playground, game room, and dining space for private parties. The number of units was the same as the original plan, except they were spread out among more buildings. The towers would have included a lot more open space, but this was what the community wanted, just like they wanted rentals. Ten months after we started construction, the project was nearly done. With the first hundred units scheduled for delivery in October, it looked like it was going to be a huge success.

Another argument in favor of moving 1000M to a rental model is that in uncertain times, not only do banks prefer rentals but buyers do too. In moments of great economic and social change, buyers would rather rent, even at higher rents, than commit to buying.

In August, a man shot by police on the South Side of Chicago set off city-wide protests—and mass looting downtown. Hundreds of people smashed storefront windows and stole merchandise along the Magnificent Mile, less than two miles north of 1000M's site. Two people were shot, thirteen police officers were injured, and more than a hundred people were arrested. There was no damage to our site or immediate impact to our project, but it reflected the state of the world at large.

To say we were in uncertain times was putting it mildly. The powder keg atmosphere in Chicago was only one piece of a much larger, and potentially frightening, picture when considering the presidential election right around the corner. Donald Trump, a merchant of chaos and greed, was running for re-election against Joe Biden. It wasn't too much to say that the fate of democracy was on the ballot in November.

We didn't know if people who had the choice were going to want to be in cities anymore. We didn't even know if people were going to want to be in this country anymore. No one could predict when or how the current turmoil was going to resolve itself. It would have been understandable if we gave up—there was a lot to overcome.

This is a business where things don't always go right. In fact, they almost never do. Damage control is always part of the mix. Although there was more damage control than usual, it is the nature of real estate. The immediate question we needed to answer was whether we wanted to invest in moving forward with a rental scenario or hold off, at least for the time being.

The whole process of completing the architectural plans and permissions for a rental option would likely cost upward of $2 to $3 million. Did it make sense to wait until circumstances improved before spending that money, or should we go for it? There's a concept called "first-mover advantage." Sometimes if you pull the trigger early, you're in better position to seize the moment if the climate improves rather than having to wait a year or more for your plans and approvals to be in place. Like so much else in real estate, the decision was a gamble.

ROB SINGER: In spring 2020, with COVID-19 in full effect, we knew that the condo plan for 1000M was dead. We needed a new plan. One option was to do nothing and simply mothball the project. The other option was to try to reinvent the project. There was a good business case for the former: do nothing and wait to see how things play out over time. Maybe the condo market would come back? We wouldn't waste money chasing different designs and business plans.

But if we were to reinvent the project, what would that mean? How much would it cost? Would the new plan fare any better than the first?

If we wanted to pursue a rental building, should we start over with a brand-new building design—with simpler, less expensive architecture? A smaller building? Would such a project even be economical, and would we be able to finance it? Ultimately, I became convinced that the best path forward would be to, in fact, reinvent the project as a rental tower.

A well-designed rental building would—one day—be economic in Chicago and would make our land valuable to us or another developer. Francis supported the idea and agreed to let me pursue the new designs. I had no idea at that time how complicated this redesign would be.

We determined early on that keeping the exterior architecture—the tower "geometry"—was mission number one. Everything would have to flow from that decision. We had paid for the architecture, we loved it, and we had approvals for it. It was our brand, and changing it would have taken years.

This meant taking on various other risks. First, this geometry was exorbitantly expensive, which is not the intuitive direction one wants to take with a rental building. It theoretically puts us at an immediate disadvantage in the rental market, where developers keep their costs as low as possible. The geometry also meant that we'd have an enormous number of units; this can also be seen as a disadvantage. We'd be designing the rental-building interiors from the outside-in, which is exactly what you're *not* supposed to do. It was difficult to make it all work: new unit layouts, new unit mix, new elevators, new amenities, new superstructure, new facade system, and new finishes.

To make it even more complicated, we realized that these changes would require new zoning approvals—which was a long process, a political process—and success was far from guaranteed. Since the new rental units would be much smaller on average than the previous condo-unit layouts, we were going to need to dramatically increase the number of units in the building in order to fill up the same building volume. So instead of our approved 430 units, we would have to get approval for 730 units; and, by the way, could we even lease that many units?

FRANCIS GREENBURGER: There was a strong argument from our partner the Karliks, who believed we had no choice but to redesign the original cantilevered glass-and-steel tower into something closer to a simple box on the basis that there was no way rents could pay for the complicated architecture. There was no arguing that the architecture we had was expensive: complicated glass, a complicated structure, and extremely tall. Rental buildings want to be as simple as they can be because that makes them as cheap as they can be.

Rob and I disagreed with their assessment, and it wasn't out of ego. When it comes to building skyscrapers, even if you wanted to do something uneconomic, albeit artistic, the market won't let you. Instead, our thinking was informed by experience. Again, we looked to 50 West where we were faced with the exact same dilemma in 2009 during the financial crisis. Although we initially considered redesigning the entire building, ultimately we decided the best thing to do was keep the exterior form and redesign the interior.

In Chicago, we were going to pay a penalty to keep the existing form Jahn designed, because it was going to cost more. However, it was more economical to pay the cost to keep the tower rather than start over. Starting over with a new tower design would require years of redesign and permits (not to mention negative press), at the end of which the city wouldn't let us scrap a beautifully designed Helmut Jahn building to put up a dumb box. The process would have taken forever and most likely been unsuccessful.

Even if we did keep the original form, the interior *would* have to be completely redesigned, just as it was with 50 West. The main concern when designing a rental building is the unit mix. If you get the size and distribution of the units wrong, you can have the best building in the world but nobody will want to rent a single apartment.

50 West

To guide us though this crucial exercise, we employed Luxury Living Chicago Realty, rental-development planners and brokers who know the market better than anybody. The company's CEO and founder, Aaron Galvin, and the director of leasing strategy, Mark Ziemke, applied their depth of knowledge of what these new rental units needed to be to the very big puzzle that was 1000M.

AARON GALVIN

Founder and CEO, Luxury Living Chicago Realty

Since 2003, I've been focused on developing a depth and breadth of knowledge about a small subset of the real estate market, Chicago luxury rentals. In that period, what has occurred most materially in the market is that any stigma around rentals in relation to condos has disappeared. In Chicago, it used to be the ultimate luxury properties were only those for sale. Now some of the rental properties are just as upscale, if not more so than many condo offerings.

In my research, I pay attention to what *doesn't* rent as much as what does. Why does a unit sit vacant? There are a variety of reasons, from too little closet space to no view to "my sofa doesn't fit."

"This is a great apartment, but . . ."

I focus on the negative to remove any objections. When people come to see our apartments, I want them to say, "I can't find anything wrong with this unit."

From seeing thousands of apartments, I have learned there are some essentials. For example, one-bedroom floor plans need to accommodate a king-size bed. If the renter has a queen-size bed, the layout becomes even better. But a lot of couples have king-size beds, and if the one bedroom in the apartment doesn't fit the king bed and frame they already own, they are going to find another apartment that will.

This project began with figuring out the right floor plans. We reverse engineered the process, starting with a deep analysis of the most accommodating and, financially, highest-performing floor plans in the city. The goal was to replicate the best floor plan for the renter while accomplishing the financial goals for the developer. We presented our findings to the developers and architects and what we considered the ideal scenario. The architects laid out the floor plate to figure out exactly how many units could fit on the entire floor. Then we dove into configuring those individual units in the most productive way. We met for several weeks, negotiating and fine-tuning until we had a plan everyone thought was the best.

Everybody had been hopeful that COVID-19 wasn't going to go on too long. We hadn't yet seen huge drops in rent prices or major changes in the way people were living in the downtown Chicago apartment market. But while we were working on the rental scenario for 1000M, we saw Chicago's luxury apartment market fall as much as 20 percent, mimicking the trends in other cities like New York.

As a company that offers not only rental strategies but also marketing plans, it was important for us to figure out the narrative of a stunning new luxury tower amidst a falling market. What were the signs we could point to that signal this development will be successful beyond the current moment?

With that in mind, there were really two things that we focused on: First, we believed the world was going to change postpandemic, especially in the way that people work. Prior to the pandemic, walking around a five-hundred-unit luxury apartment building on a Wednesday afternoon, you might see twenty-five residents working in a common space. By 2020, you see double that number—with time limits of an hour and a half and reservation systems. People were working from home, and we didn't think that would go away when the threat of COVID-19 subsided. Some industries, we presumed, may adapt to working remotely on a permanent basis. But even for those who go back to the office, there would be a huge element of working from home beyond the pandemic—whether to space people out in offices through shifts or change the work-life balance.

That vision of the future came into play when designing the units, ensuring there were additional spaces for working even in the studio units. More importantly, however, was how it impacted the amenities. We put the coworking spaces front and center to create an unparalleled sense of community in 1000M.

Our other area of focus was what we saw happening with "ultra" luxury rentals in Chicago. They didn't seem to face the same challenges as the rest of the market. If the market had dropped 20 percent, those projects, of which we represented a few, either maintained or even increased rents. While the world was falling apart, Mark and I were renting one-bedroom apartments for $4,000 a month in downtown Chicago. So the developer's initial instinct to keep 1000M on the high-end was proving correct, even during the worst of times of a pandemic.

MARK ZIEMKE
Director of Leasing Strategy,
Luxury Living Chicago Realty

Since 2016, twenty thousand new luxury apartments had been built in downtown Chicago, but there was a very small handful of buildings we considered ultra-luxury rental buildings on the level of 1000M. To make this conversation a success, the rental needed to keep almost the exact same quality in the lobby and of the finishes, amenities, and architecture—everything that the developers were going to do in condos.

FRANCIS GREENBURGER: In the final analysis, Luxury Living's recommendation turned 1000M from a 506-unit condo tower to one of 738 rentals. Their presentations on the unit-mix and floor-plan financials was thoroughly researched and persuasive, but would the geometry of Jahn's carefully crafted skyscraper accommodate such a demand? As it turned out, the form of the building was not the most monumental task ahead of the architectural team.

PHIL CASTILLO
Executive Vice President, JAHN

Francis really didn't want to change the design of the building—same with Helmut. To the untrained eye, the building looks the same. 1000M still has a delineation between the base and the tower, that we refer to as the cut, although we simplified some parts. On the facade there were various kinds of profile panels that became all flat panels. This facade treatment is also now only on the Michigan Avenue side, like the historic buildings along Michigan Avenue, which had a nicer facade in the front and plainer ones on the other sides. The tower

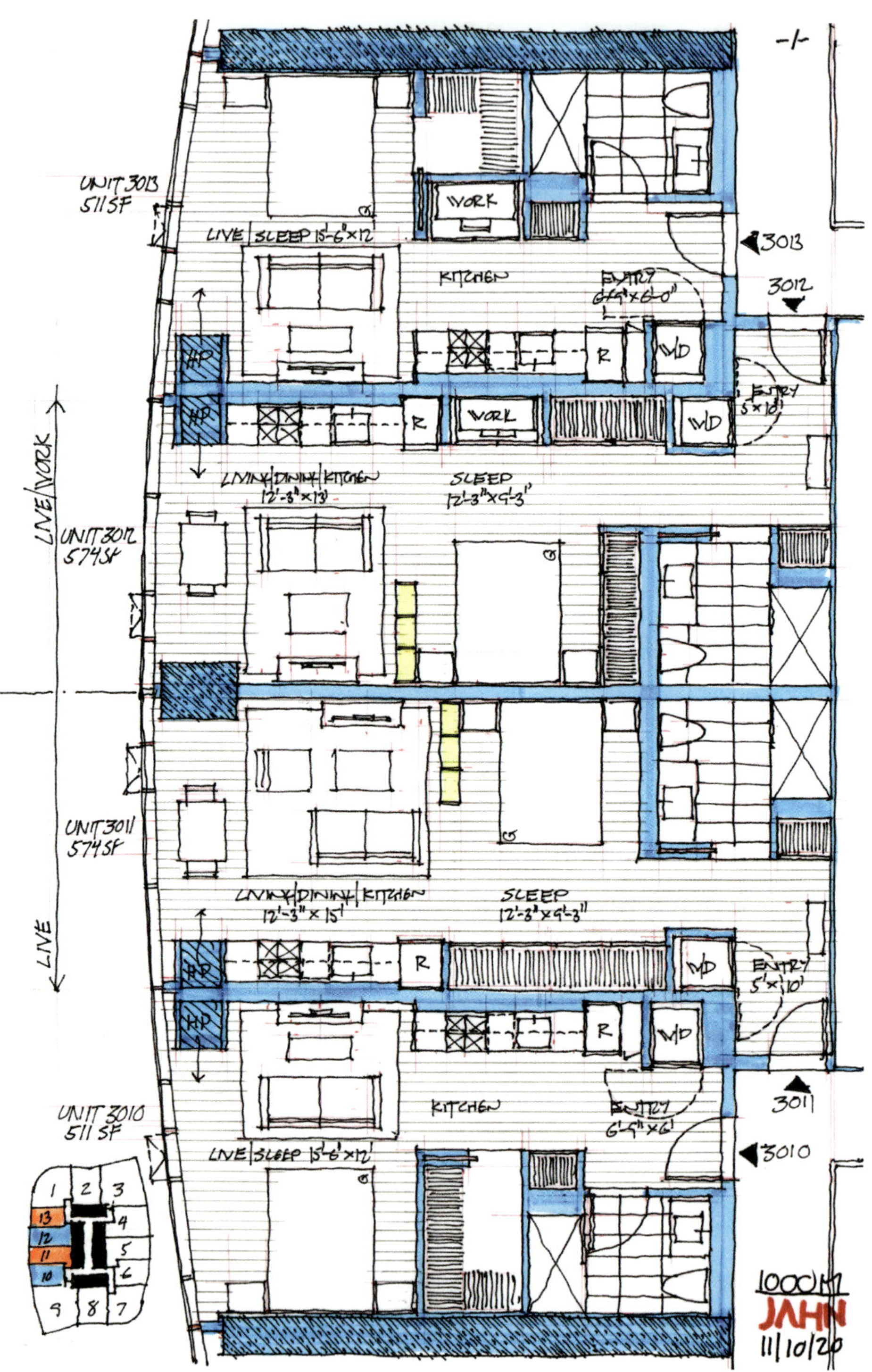

Sketches of the apartment layouts in development, November 2020

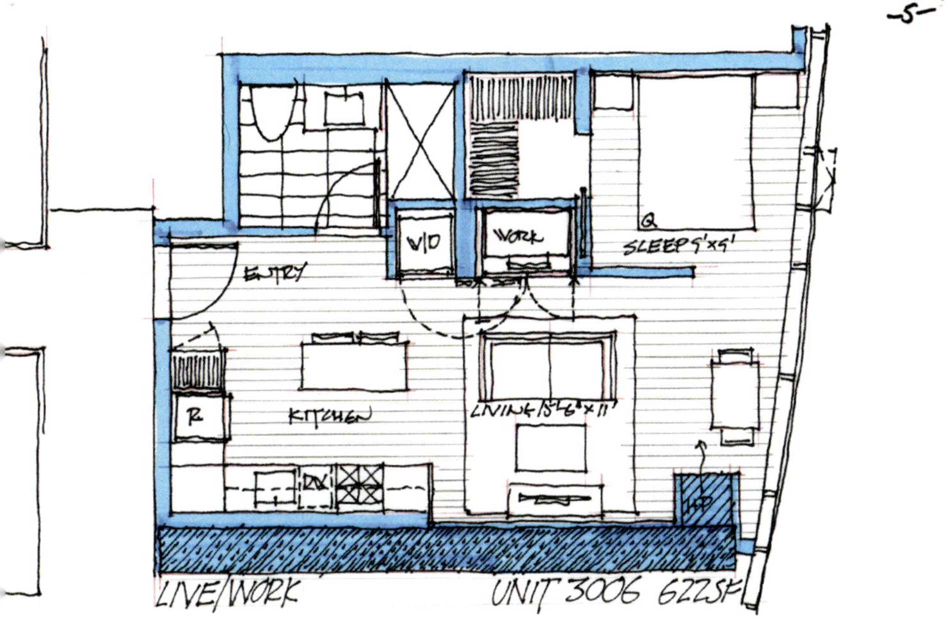

W/D
SLEEP 9'x9'
ENTRY
KITCHEN
R
DW
LIVING 15'6"x11'
HP
LIVE
UNIT 3006 622SF

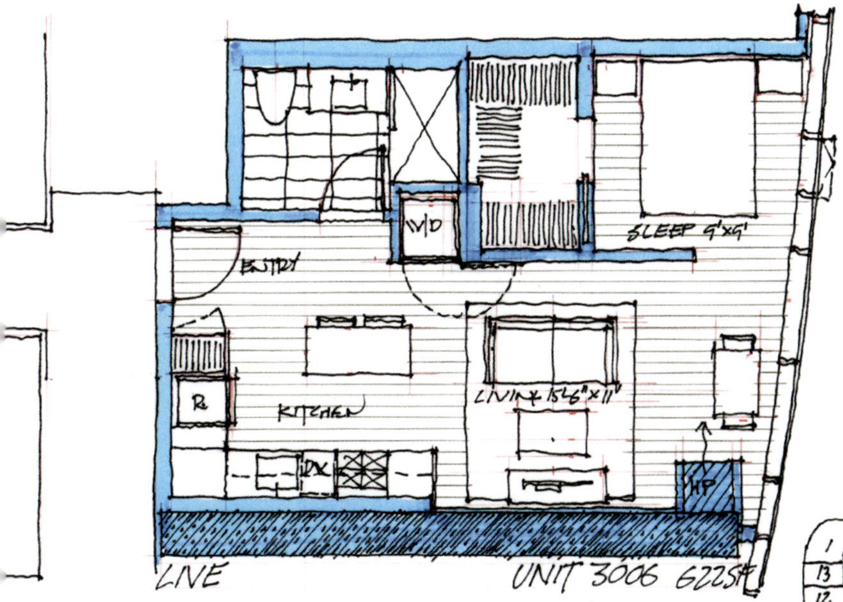

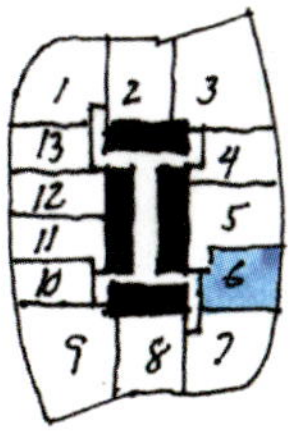

1 2 3
13 4
12 5
11
10 6
9 8 7

shrunk a bit as well; it went from 830 to 800 feet tall because the ceiling heights in rental buildings are typically reduced.

A large part of the effort was working on the apartments. When Luxury Living gave us a whole stack of apartments they thought were good, you just get your head screwed on right that day and start working on it. We had meetings with Luxury Living and Rob every week for a month and a half or so to get these units to a place where they wanted to be. Half of the apartments—which are much smaller than the original condos—are studios. Floors twenty-two through fifty-four have essentially thirteen units per floor, where they were originally supposed to have seven or eight, in varying mixes of studios, studio convertibles, one bedrooms, and a couple of two bedrooms.

The interesting problem we started getting into was adapting a unit like a studio convertible to a work-live unit since so much of the population was working from home and probably would continue to do so. What do you have to change without altering the kitchen, plumbing, and anything else crucial? We came up with kind of a clever little plan where some of the closet space disappears and becomes a home office. Two big closet doors about three feet wide that open up from the middle slide back to reveal a place for a desk, bookshelves, and a printer. Then, when you're done, you can just close it up and you don't have to look at it.

The tower has much more amenity space now than it had before—a rental building needs much more amenity space than a condo building. We planned for 1000M to have eighty thousand square feet of amenity space. Before, there was essentially just one floor of amenities. In the rental scenario, that grew to six floors. Instead of one outdoor pool, it now had an indoor and outdoor pool. In Chicago, rental amenities are a big deal. They're a much bigger deal in rental buildings such as NEMA or Wolf Point. NEMA, Rafael Viñoly's seventy-six-story tower over Grant Park, has a half-court basketball court. There was no question that people still wanted amenities, even post-COVID-19.

The increase of amenity space, which we programmed with Rob and Jordan, helped the building a lot. With multiple floors to play with, we were able to separate the space between active and more sedentary activities. After we moved the fitness area to another floor, the children's room was no longer shoved in a corner, as it was in the previous building, and the game room was no longer part of the lounge.

The two floors at the cut in the building where the glass tower separates from the base became a separate amenity level just for those renting the skyline units. Those were larger apartments on floors fifty-five to sixty-nine, where the density dropped to seven units per floor and, on the three penthouse floors, four units per floor. These more expensive rentals had access to the more private amenity level with four little terraces, a smaller workout room, a couple of dining rooms, and, of course, a coworking space.

LAURA LEE MCALLISTER

Director of Architecture, Kara Mann Design

In some places we did keep the thought and the design from the original condo building because it would successfully transfer to the rental. That included the main lounge space for all residents on the eighth floor. Overlooking the park, it's a cozy library-esque space with a fireplace. The winter garden stayed as well. The seventy-third-floor indoor-outdoor space on the corner at the very top of the building was always a killer space. On a cold, miserable Chicago night, you can sit, warmed by one of the fireplaces and an interior filled with green plants, and look over the city, lake, and park. Who wouldn't want to have a dinner party up there?

One of the biggest challenges for us in the move from condo to rental was the budget. The team had seen very high-end work from us for years, and it had to remain that way—but how could we do it at a much more affordable price point?

We sourced materials from very different places, less expensive vendors, commercial-grade fabrics. We have that mix in our wheelhouse because we do rental buildings, hotels, and the like. It was difficult, however, to make that shift in our brains when we had spent so long solving problems one way. Our interior design needed to look as good but also be budget friendly.

We searched for products that looked similar but were one step down in the material finish, like Luxury Vinyl Tile (LVT) over real wood, human-made stone over marble, or laminate cabinet faces rather than real wood veneer. The hardest area to match was flooring in general. We scoured a wide range of samples in search of a material that looked like and had the texture of real wood. It took a lot of time and effort, but we found a product by Shaw Contract with some grit to it (meaning it didn't look too perfect or polished) and a high-end finish. Carpeting was another element, which was difficult to find something that didn't feel cheap. The rents in this building weren't going to be cheap, so they required carpets that felt soft and lovely—at the lowest price.

In the condos, we used human-made surface for the kitchen and bath countertops so they would be durable. We sweated, in a labor of love, to find the perfect stone for the original condos. We must have gone through fifty to seventy stones with Francis. It was a little painful to think back on all the effort we put into that, only to have it not happen.

FRANCIS GREENBURGER: All of a sudden all these things started to look really, really good. The architecture and design teams, working hand-in-hand with Luxury Living, Rob, and Jordan, were presenting plans for a stunning tower. The numbers were coming back and construction costs were in line to deliver a state of the art building in 2024,

Dining room, floor 73

when the world was going to be a lot better. Except now there was a new problem—there always is.

The condos had averaged about 1,300 square feet per unit while the rental apartments averaged about 800 square feet. What that meant was in order to fill up the same volume of Jahn's original building, we had to create many more units. Our zoning allowed for 505 units, but to properly plan the rental we needed 740 units. We had two choices: get approval from the city for significantly more units, or start all over again. It seemed as though we really only had one choice.

We met with the lawyers and made a plan to return to the city to ask for more units. In mid-September, the team made a presentation to the planning board of Chicago, which was receptive to our argument. City planning has to agree with the basic plan as well as approve of the technical aspects, but in order for the board to start processing the application, we needed the alderman's blessing.

Sophia King, the alderman of the ward in which 1000M sat, had been a big champion of the project. She spoke at our condo groundbreaking, lauding the tower's beauty, all the jobs it was going to create, and the taxes it would generate. So we called her back with the news that we couldn't move forward with the condo, but we were prepared to risk an enormous amount of money to construct a rental building. All we required was for her to support the additional units, which the city supported. Unfortunately, September turned into October, then November rolled around, and she was still ignoring us. *That* was a problem, and it wasn't our only one.

When we began the original condo development, the first part of the overall financing strategy was a $27.5 million loan Goldman Sachs gave us against the land to start the construction phase and complete a lot of the design work. The idea was that the investment bank would follow up the land loan we closed in November 2019 with the construction loan. It was a one-year loan, because we planned to roll it into the full loan before it came to term. When, in summer 2020, Goldman turned around and announced it wasn't going to make the construction loan, the first thing I did was request that they extend the land loan. "I don't want to get a notice in November saying I owe you $28 million," I explained. A reasonable request, so they agreed to an eighteen-month extension of the loan to give us a little breathing room—simple enough. Again, though, nothing is simple in skyscraper development.

Goldman really dragged its feet on the paperwork. The documents had been approved by their lawyers, then by ours. We signed them and returned them for the bank to sign. But weeks turned into months, and still the papers weren't signed. When I called Rod to find out what the holdup was, he explained they had a shortage of personnel. They were down a lot of financial analysts who do all the complex underwriting and were having trouble hiring new ones. I was surprised that one of the biggest investment banks in the world was short staffed. But Rod assured me there wasn't going to be a problem with the loan or the terms.

By November, when Goldman *still* hadn't processed the paperwork, we did start to worry. Rob was calling them every day asking for its status until November 7, when we heard back. Bad news: A committee decided they needed us to reappraise the land as a check on the value before they would agree to close on the loan extension. If the appraised value of the land comes back lower, they wanted me to write Goldman a check for the difference.

"Are you kidding me?" Rob went ballistic. "We've been working on this for five months, and two and a half weeks from the loan being due you're telling us that you're going to hire a brand new appraiser to go in and figure out what the value of this vacant land is in the middle of an international pandemic?"

The anxiety didn't have to do with the logistics of another appraiser coming out to look at the pit that was currently 1000M—it was what number they were going to come up with. As Rob said, "It's not worth anything. There's no buyer for it!"

SLOPE
SLOPE
SLOPE
1000M

Chapter 8

POLITICS, WINTER 2020

1000M Hits an Unexpected Obstacle

FRANCIS GREENBURGER: Rob called me in a panic. He was preparing his presentations for Goldman and their appraiser in advance of the appraisal of 1000M, and it wasn't looking good. Rob works with appraisers frequently to explain the vision of a project and why the land is worth so much. So when he told me he was worried, I trusted that he had good reason to be.

In winter 2020, it seemed like everyone had moved out of the cities: Chicago, New York, everywhere. According to the pundits and experts, cities were done. Occupancy rates that had been at 95 percent had plummeted to around 65 percent. Rents dropped to half their value. This data threw the whole theory of new construction in a city into question.

So when the appraisers returned after Rob's presentation with the assessment that the land's value was *higher* than in 2019, we were amazed. They wrote a glowing report about how 1000M was going to be one of the best buildings in the history of Chicago, worth half a billion dollars. The appraisers approved of our design and business plan. To them, everything we had done made perfect sense.

We weren't expecting to have to do an appraisal and railed against it, afraid that it was going to come back zero. But in a twist, it came back a plus. Not only did the value increase, but we had a third-party review that we wouldn't have had otherwise for many months, one that made the process of getting the capital stack together much easier.

We threw it right in Goldman's face. Their own appraisal showed that our plan was good and the coverage even better. According to the appraisal, the project would be worth $470 million upon completion. Meanwhile, we were only asking them to make a construction loan of $310 million. They renewed and extended the land loan, but now we were pushing them to use the same report to reconsider the construction loan. It showed that we had a solid business plan for 1000M, which met the underwriting criteria the investment bank needed to make the loan without getting hurt.

As the loan committee considered our request, everything started to feel oddly positive. We were sitting on an unparalleled rental project with the foundation built, just ready to go vertical. The pandemic might still be raging, but Goldman wasn't going anywhere. It was still a bank that needed to do business and meet its revenue targets. In order to do that, it had to make loans. And if they were going to loan to anyone, 1000M was a better-than-average risk. Construction loans are safer for banks than loans on apartment or office buildings because they have more equity behind them and higher rates. With a construction, the bank makes their money off the interest during the period of building, all of which is built into the budget. There is no risk that they are not going to get paid (barring the developer going bankrupt, which I did not plan on doing.) Only two weeks earlier we were fighting for our life, and now we were on top of the world. It was enough to give you whiplash.

We knew better than to get comfortable. As I said, skyscraper development is an endless series of violent motions forward and backward. There is no coasting. But no sooner than we received the positive appraisal we encountered a new bump in the road. In December, we finally heard back from the alderman whose office Rob had been calling for months to get approval for more units to accommodate the rental tower.

Although Alderman Sophia King was generally supportive of 1000M, her actions became more difficult and demanding once we asked her to approve these additional units—even though the overall size of the building was identical to what had already been approved. In my mind, the fact that we were going to have more apartments available as rentals—as opposed to larger apartments that could only be afforded by higher income families—seemed like a political win. But Alderman King did not seem to buy into this logic. Instead, she made it clear that she would only approve the project if we met certain standards, which were tougher than the city's normal standards. These included increasing the number of parking spaces and affordable housing units as well as the percentage of contractors who met the city's MBE (minority business enterprise) and WBE (women business enterprise) standards. The issues were complicated because what she was asking for, beyond the city standards, would have meant an increase in the costs of the building, which were already at the breaking point. I tried to explain that if we could not achieve a manageable budget, the banks would not approve the plan, and the building would not be built. The affordable and conventional housing that it would provide would not come into existence. Whether this was persuasive to her or not, I do not know. I tried to engage her in a conversation about these issues, but she was very difficult to communicate with, often not setting up meetings. She also requested that we host community meetings but made it very difficult to schedule them. Rob was plowing ahead with the capital stack and the design, which continued to cost money to advance. What if the alderman didn't come around? There were many, many millions of dollars in the balance.

JACK GEORGE

Counsel, Akerman LLP

I don't buy into the idea that politics dictates the process. Don't get me wrong: There are some instances where it takes you longer to get meetings, but I've never met anyone who is going to stop a building because of a political issue—and I've been doing this for fifty years and have never had a development I worked on that hasn't been approved. Developments are either passed or not passed based on their merit, no matter the mayor or alderman. You can't tie your success to who is in office. There may be some stumbling blocks but that's not what stops them from passing if they don't.

When things are not getting done fast enough for the developer, or the alderman is asking too much, it is my job to be an advocate for my client as well as make the client understand what is going on. My job is not to get aggravated, to stay levelheaded. I completely understand a developer. If I were putting millions of dollars at risk and someone was giving me the business, I would be frustrated too. At the same time, my job is to get the project done. While I take the position of my client, I don't take on the emotion. Because if I called up an alderman and said, "Hey, you creep," they would tell me to take a hike.

Calling people names or telling them they aren't doing their jobs is not the way to resolve whatever issue is holding up the project. Instead, that's a surefire way to get your development bumped to the bottom of the to-do pile. I had a client call me last night at 8:30 p.m., extremely angry. I had to convince him that if we could just wait a little bit longer, we would be able to resolve the issue. I majored in psychology, economics, and philosophy at Xavier University—and some of my studies have helped me over the years. I love the give-and-take of zoning, working with different personalities; I find it exciting.

FRANCIS GREENBURGER: I trusted Jack George, but at the end of the day we still didn't have the alderman's approval. Meanwhile I agreed with Rob to release a couple hundred thousand dollars to pay the architect and interior designer. That was enough for them to work for the next thirty days, when hopefully the issue would be resolved.

ROB SINGER: In winter 2020 going into early 2021, working on 1000M continued to be stressful. We were slugging it out with the city, trying to finalize approvals to increase our unit count up to 738. At this point we believed that we had a good base design for the rental tower—if we could get the new units approved.

We struggled to make progress on financing because lenders were still skittish about lending on large construction projects. But then, by February 2021, things began to change. Aaron and Mark from Luxury Living called me: "Something weird happened," they said. "We just rented fifty apartments at a luxury building this week." Meanwhile, they hadn't rented an apartment in about a year. Was this the start of a new market? Or just an aberration?

The COVID-19 vaccines had been approved and were rolling out. People were moving back to cities. Some predicted it was probably just a blip, but as more time went by, it was clear this was no blip.

People in cities, including Chicago, wanted to rent apartments again—and they were willing to pay a lot of money for new apartments that were good. Around this time, we reengaged with Goldman to alert them to the changing market dynamic. The luxury rental market was coming back, and there was not much new supply under construction or that was planned. We saw a window of opportunity opening. Interest rates and inflation had not yet begun to spike.

Goldman was happy to reengage. They hired a local consultant to review our new project, plans, and economic feasibility. The study showed that the 1000M rental design, timing, and business plan were economical. It seemed we finally had the right project at the right moment.

FRANCIS GREENBURGER: Goldman Sachs was convinced enough by the shift in the urban rental market that it started up the process again of considering the loan for 1000M. My contact claimed the president of the bank blessed the deal, which would obviously be a very good sign but far from the end of the road. Now the loan request had to go through the bank's separate asset allocation department. At least it was progress—maybe.

We had made progress on the political front as well. It took far longer than I had anticipated or hoped, but by March we had made a deal with the alderman. I think sometimes elected officials are not as sensitive to the real-world economic constraints that all nongovernmental projects and businesses have to operate within in order to accomplish significant projects. The relationship between government and the private sector is best served when a cooperative private-public relationship is established that respects the needs of each party.

The alderman gave us two hours' notice before she held a community meeting over Zoom. Although a number of her constituents wanted to know why she scheduled it so last minute, the meeting wasn't hostile. A development expert was invited, who asked a series of technical questions, all of which we answered.

Now there was a whole process to go through until we could claim the re-entitlement for the extra units. Once we were able to satisfy all the different stakeholders' requirements, then the matter would be placed onto the agenda of the Chicago Plan Commission made up of twenty people with different specialties, including architects, land planners, aldermen, and other community members appointed by the mayor's office. As the recommending body, they hold a public meeting where anyone from the public can attend and speak in favor of or against a project. If the plan commission votes to recommend the land use or zoning variance, it has to go to the zoning committee of the city council for another open meeting. This committee is made up of fifteen to twenty aldermen to whom you put your case again. Lastly, it has to go to the full Chicago City Council for a vote.

We were winding our way through this lengthy bureaucratic maze when an unthinkable tragedy struck: Helmut Jahn, an avid cyclist, was riding his bicycle about fifty-five miles west of Chicago on Saturday, May 8, when he was killed in a traffic accident. He had been riding on a street in Campton Hills—not far from the horse farm where he lived in St. Charles, Illinois—when he was struck by two cars headed in opposite directions after he went through a stop sign at an intersection. Helmut, pronounced dead at the scene of the accident, was eighty-one.

He still had so much life left in him. Obviously Helmut's unexpected death was an immeasurable loss for his wife, Deborah, and son, Evan. But a man, whose personality, drive, and force of creativity were as large as the starchitect's, leaves behind a large gaping hole when he is gone.

Other than his family, perhaps no one stood to suffer Helmut's absence more than Phil Castillo, the architect's right hand, most trusted colleague, and employee of the firm since 1979. The shock he experienced when he picked up the phone at 10:30 p.m., uncharacteristically late for a Sunday evening, only to learn that his longtime partner was dead.

PHIL CASTILLO: We were close. I'd been with him forever, as most people know. But we were doing what I believe he would have wanted us to do: design buildings. Fortunately, we were busy, which helped, and happy that 1000M was moving forward. It was an important project for us, and it still is; it's probably one of the more important projects. 1000M is certainly the tallest building we've designed in Chicago, and the tallest building we've designed in a long time apart from the seventy-story CitySpire in Midtown, New York, and Leatop Plaza in Guangzhou. So for us, it's important.

Philip and Helmut

FRANCIS GREENBURGER: Phil showed a characteristic strong and practical resolve in keeping the firm humming as it continued to go through the motions of a world-class architecture firm with projects on the docket—even as its staff absorbed the trauma of unexpectedly losing its leader. I was hardly surprised by Phil's steely Midwestern resolve. He has always had a cool head and firm hand, no matter the chaos or pressure of the circumstances. Still, I admired how he kept the group plugging away at 1000M even as he mourned his longtime partner.

While I had nowhere near the relationship with Helmut that Phil did, I, too, grieved his loss. I had worked with Helmut on three projects, including the completed 50 West Street and our most recent collaboration, 1000M, whose fate was still yet to be determined.

I had been warned before I started working with Helmut fifteen years ago that he could be "difficult," but that was not my experience. Instead, I found someone who was always relentless in his willingness to perfect and improve his work.

Like all successful architects, Helmut had a strong ego and clear vision. But he was an extraordinarily keen listener. Helmut made it his practice to present his evolving ideas to me. He constantly took me through the evolution of his thinking while soliciting additional feedback. If you offered meaningful direction and suggestions for continued improvement, Helmut would seize on it and return in a few days with multiple new designs. This work often represented endless late-night sketches that Helmut would immediately send off to his team while flying around the world, expecting them to perfect these sketches and model them by the time he returned home the next day.

As we waited for the city to consider our application for the redesigned building, it was my greatest aspiration that 1000M one day stand as a legacy example of Helmut and Phil's best work—and serve as an icon of Chicago's renewal as we hopefully came out of this very challenging pandemic.

I was far from the only person who saw this tower overlooking the lake in a new light after Helmut's death. When the Chicago Plan Commission approved our request for zoning changes in mid-June, the lead of the *Chicago Tribune* article reflected this sentiment: "Helmut Jahn's tallest Chicago building could soon resume construction after a series of setbacks that stalled the South Michigan Avenue project, including the famed architect's death in a May cycling accident."

The article covered the news that 1000M had shifted from a condo tower to a luxury rental. The changes, the *Tribune* reported, included reducing the height of the building 27 feet for a total of 805 feet. In the redesign, 1000M had lost one floor to become seventy-three stories. However, the apartments had grown by 217 units, for a total 738 apartments. "At the latest, resumption of construction is expected by next spring, Time Equities Chairman and CEO Francis Greenburger said in emailed statement," the article states. "The statement did not provide details about financing."

There was a good reason for that: We still didn't have a commitment from Goldman Sachs about the construction loan. At that point, there was no other bank who could come in and take their place. All we could hope—as the zoning committee approved the plans on June 22, 2021, and the Chicago City Council voted unanimously for it a week or two later—that the investment bank would be reassured by the city's full support of 1000M.

Chapter 9

REALLY SMART PEOPLE, FALL 2021

Closing the Construction Financing, Despite Everything

FRANCIS GREENBURGER: The architects and interior designers were working away, filing the necessary documentation and moving the plans forward in a timely manner, even though we were still scrambling for financing on all fronts. So cutting a check for $420,000 to the entire design team in March 2021 didn't feel particularly good, especially since rising interest rates were making the amount we had to borrow to build 1000M more expensive by the day.

As the threat of the pandemic ebbed and flowed, the world had decided to move ahead. Starting in January, we began to see a reversal of the trend of people leaving cities. In New York City, where for the previous nine months rentals were all in the red, we had a positive leasing in our apartments. 50 West, which never had big vacancies, had a 7 percent vacancy for the first year of the pandemic. Now it was 100 percent full.

Our new development in West Palm Beach was a phenomenal success beyond anyone's possible expectation on every level. About a year ahead of schedule, we were leasing apartments at a rate of three times what we had predicted. CasaMara was the hottest property in one of the hottest markets in the country.

Casa Mara

That was important to Chicago for a number of reasons. It was another demonstration of the competency of Time Equities that I never doubted. We now had a successful track record we could point to in new developments. I was also able to negotiate a very favorable permanent loan for CasaMara. (At the end of a construction project, you pay off the construction lender and get a permanent loan for the long-term.) To pay off CasaMara's $75-million construction loan, we got a permanent mortgage for $105 million—leaving us with $30 million extra, which we could use for the equity portion of 1000M.

Still, Goldman Sachs continued to drag its feet on the construction loan for 1000M. After a winter of the investment bank continually promising to put the deal into the committee "next week," they finally did in April—and the result was very disheartening. They still wanted to lead the loan, but instead of giving us the entire $300 million of the total $430 million project, Goldman was only willing to put in $200 million. And they wanted us to find the partners willing to loan the last $100 million that we needed. So we were back to slugging it out. I knew my team would eventually put together a group of four or five banks to chip in the money, but it was going to be an enormous amount of work and a very "hard lift."

After Rob had spent the winter pushing Goldman for a decision—and in return getting only more questions for which he had to provide their analysts more and more data—he and the rest of the team at TEI geared up for the process of identifying and approaching endless banks about building a new tower during a pandemic. We all knew we were facing an uphill battle when, out of the blue, Rod from Goldman Sachs had an idea. He knew someone at Deutsche Bank who might be interested in the deal.

Anthony Valvo, who had worked with Rod at Citigroup during the '90s, was now a regional managing director for Deutsche Bank's Wealth Management Private Banking office. While a loan of this size was not exceedingly rare for his group, a construction loan of any size, let alone one for $100 million, was very uncommon for the leading German bank. Private banking is often leery of construction because in general it comes with significantly more risk than a traditional real estate deal involving existing properties. The uncertainty created when all you have is dirt and a vision is simply too high.

The result was that despite making billions of dollars' worth of loans annually, Deutsche Bank's wealth-management side might look at one construction loan a year. The odds weren't great, but in this case the bank might be willing to make an exception. Before arriving at his current job, Anthony worked at Barclays where he was assigned to TEI's account. It was an incredible stroke of luck that he was not only friendly with Rod but also knew my team and me.

When everything having to do with 1000M had been difficult, hard to the point of biblical, suddenly we had a guy who seemed to be saying, "Yeah, I'll do the $100 million. No problem." It wasn't quite as easy as that. The bank would still have to do its lengthy due diligence, but by the fall we had a commitment letter from Goldman Sachs for $200 million and were moving in the right direction with Deutsche Bank for the last $100 million.

There was just one issue, and it wasn't a small one. We still didn't have the full $130 million in equity we were responsible for. The $75 million mezzanine loan that would complete the "capital stack" was not much further along than when we had started on it a few years earlier.

A TEI analyst researched the viability of a CPACE loan. The financing vehicle—which stands for "commercial property-assessed clean energy" and is only available in states that enacted the legislation—seeks to promote energy efficiency and renewable energy through loans for the green aspects of building. Although the loan can be private or government money, the loan is repaid via an assessment on the property tax. In addition to all the red tape involved in the loan, the loan structure mostly didn't work—at least not for new construction. However, a CPACE lender had promised us that this time would be different; they would subordinate their loan to the construction loan until the project was complete and the construction lender paid back.

I was going to use the CPACE loan as part of the capital stack instead of preferred equity and pay it off when the building was finished. The rate they were offering of 5.5 percent was better than the 15 or 20 percent from a hedge fund for a typical mezzanine loan. But it was higher than the 3 percent I hoped to get for a permanent first mortgage when the building was completed and leased up. The CPACE lawyers insisted that they be in the most senior position in the capital stack, contrary to what the CPACE lenders, who were very anxious to get the deal done, had told me—I imagine because they didn't have a lot of other takers. They were basically selling doughnuts for almost twice the price than you could get them across the street, so why would anybody buy them?

I was already working on another solution for the mezzanine loan, a concept so innovative some people thought it was nuts. A 1031 exchange allows those who sell their real estate to delay the taxes on the sale if they reinvest the proceeds into another piece of real estate. There was a great deal of demand from 1031 tax-motivated investors, because the value of residential properties typically favored in these transactions went up during the pandemic. There were a lot of people with tax pressure, but the deals to alleviate it were harder than ever to find.

There was, however, one major snag. The law says you and the investor must conclude the exchange with 180 days of the original sale. How could you invest in a building that was going to take three years to build? The pieces didn't fit.

I like problems and puzzles. Even better, I like solving them. I came up with a straightforward idea to address the issue. If we sold the land for 1000M to a new entity made up of as many owners as needed, we could lease the land from this group to develop the building. The 1031 exchange allows for people to buy investment properties. We would pay rent that effectively gave the land an income stream from day one. When I floated the idea to my team, I suggested the initial rate of 5 percent on the land that we'd sell for $70 million. That meant our lease payments to the landowners would be $3.5 million a year. Once the building was completed, the rate of return would increase.

We conferred with the banks; there was no reason we couldn't move ahead. However, there was one caveat. If the banks foreclosed on us, they were taking the land from these landowners as well. The construction loan we were hoping to get from Goldman and Deutsche Bank was still senior to everything else.

Imagine a mom-and-pop operation sells their building for $4 million and reinvests the money in the land. They would earn $200,000 a year but would also be subject to getting totally wiped out if the developer defaulted. Was 5 percent, increasing later, enough incentive to risk losing everything?

Initially, the team at TEI pushed back. They thought the return was too low. They argued that nobody was going to subject themselves to a $300-million construction loan with development risk for that kind of money. My argument was, yes, they were taking on more risk than they normally would, but they would also get into an A-plus property that they normally wouldn't have access to.

DAVID BECKER

Managing Director, Equity Division, Time Equities

I had been telling Francis that there was a lot of overflow from 1031-exchange demand. "There's this equity for the taking," I said, "and we just don't have the projects for it." Meanwhile we were struggling to finance this major development deal. The mezzanine offers we had were at a price so high that they severely impacted the project because the cost of the money was too burdensome.

Then I got an email from Francis. Out of nowhere, he merged the demand that we had on the 1031 side and the need for financing for this new development project. In typical Time Equities' fashion—true to Francis—we always do something different and unique, and we do what works for us and others. This was no different. Really, what it was, was a paper transaction for a ground lease structured as a mezzanine loan.

Conceptually, I liked the idea but originally, I didn't like the initial return he wanted to offer—which hovered around 5 percent—to investors, who were used to getting more. "Francis, this is not like a core Class A multifamily. This is a development deal, which we all know carries more risk. The market knows it carries more risk."

"This is what the project can withstand," he said.

"Well, I just don't think this is going to take."

My job at Time Equities is to be the liaison with the investors and to provide a high-quality offering. I knew the project was going to be high quality, but what were we actually offering the investors?

My team and I hit the market, actively speaking to everyone we could. The range we approached included some generational investors, looking to give their beneficiaries something of substance. A beautiful new building that would last way past their lifetime might be it. Other investors considered the deal because they knew there was a very reputable sponsor and development team behind 1000M. While we might not be offering the highest return, Francis gave people a lot of security. Others were looking at the option solely because they had nowhere else to go.

What we had going against us was that our everyday investors shied away from Chicago. At the time we were going out for the money, there could not have been worse press about crime in the city. Social issues weren't the only obstacle—people were starting to get hyper focused on the state taxes of real estate investment locations. This wasn't a Sunbelt deal. Illinois didn't have the favorable tax structures of Florida or Texas.

Because this was a big project, I did not want to come up short. Still, I needed to manage Francis's expectations so that he could look at other options in case no one was home on this. When I had a conversation with an investor with a $20 million exchange who liked the creativity of the deal but not the long-term time frame, I immediately went to Francis for more creativity. "Could we structure something to give him an option to exit?" I asked him.

Even if we didn't hold investors long term, I started to think 1000M could serve as a good warehouse for someone to essentially sit in it for two or three years. Francis wasn't sure, but I was circling $20 million. We came up with a concept of essentially giving investors what I regarded as a Time Equities bond for five years, which would enable us to build the building and stabilize it (which means lease it.) Then when we moved to permanent financing, the 1031 investors would have an option to exit. For five years, they'd get a current 5 percent annual return. Then they'd have a choice either to exit or roll into the permanent transaction, in which case they would get roughly about half of their original capital back in a way that the tax was deferred and a growing senior equity return on the balance of their investment.

The investor committed the $20 million and then, right there, I knew we could get home. No one likes to be the first person in the pool. But when you go to people and say, "I have a $72 million offer and $20 million already raised," they start jumping in. One broker-dealer signs off and then another hears about that and signs off next. There is a momentum effect, like with any market. In the end, all the money was raised inside six months.

The entire way Francis conceived of this exchange offering was hugely beneficial, but we also hit the market at the right time. As much as 5 percent didn't seem attractive to me for what I thought these people wanted, there was nowhere else to go for any sort of yield. Plus, it was a 5 percent *net* to the investors, because we covered financial advisor or brokerage costs. Paying placement costs was nothing compared to what we were saving.

According to my estimates, the mezzanine money brought together in this way cost somewhere between one-third to one-half, less than what the preferred equity lenders would have priced. It was 10 percent less interest per year—on a $70 million loan, over the five-year offering, that would be a savings of $35 million in interest. That's serious money.

FRANCIS GREENBURGER: As often happens, what once seemed impossible quickly turned into inevitable with just a moment of success.

It reminded me of my early days as a literary agent (which is a whole other story). I was trying to sell the first novel by an unknown writer. I liked the manuscript, a faced-paced mystery, but I couldn't find an editor who agreed with me. After getting dozens of rejections, including from an entry-level editor at Little, Brown and Company, I decided to send the manuscript to a more senior editor I knew at the same imprint. He agreed to take a chance on *The Thomas Berryman Number*, the first book by the one-man writing juggernaut James Patterson, and the rest is publishing history. The lesson is that the first no—or even the first twenty—is not necessarily the last word.

September 24, 2021 was an important day. We closed the first phase of the infamous land-lease deal, the idea that everyone thought was idiotic but had the potential to become an extremely powerful tool. While closing on $72 million, David was already finalizing the next phase. Financial innovation has always been a part of my real estate career.

Back in my early days of turning rentals in New York City into co-ops and condos—the cornerstone of my real estate career that allowed me to diversify my company far beyond the scope of apartment sales—I invented a special financing instrument. During the process of turning a Bing & Bing building in the West Village into condos, the tenants committee approached us with a problem. The committee of the prewar building, consisting of mainly studios and one bedrooms, thought the price we were offering to purchase the units was fair. However, there were certain tenants who couldn't afford the asking price even if it was "fair," and the committee were understandably reluctant to leave them out of the deal. I came back with a proposed solution: Tenants could name their own price, but they would only own a proportional share of the apartment. So if a tenant wanted to pay half of the asking price, they owned half the value of the apartment. That meant when they went to sell it, instead of getting the whole sale price, they would get half of the profits. In the fullness of time, I've received checks for $1.25 million for apartments we turned into condos in 1986. I also enfranchised multiple potential buyers who otherwise would have not been able to get into the market.

Similarly, we pushed the envelope to come up with a deal for the 1000M's mezzanine loan that was a win-win. We had searched the world for the preferred equity, and there was just nobody home at a reasonable cost. Then up popped this market of people with tax pressure. By offering them 5 percent interest growing and a senior ownership position in the building, for what a hedge fund was charging 15 or 20 percent, we got the loan for a much better rate, and the investors solved their tax issues.

Now that we had the mezzanine loan and developer equity commitments filled, we began intensely pushing for Goldman to fulfill its commitment. With every passing day, closing the construction loan became more urgent.

We had been moving ahead with construction plans. Over the summer, we had authorized Jahn to complete all the construction documents so that we could get the final construction bid. Construction management firms submit a guaranteed maximum price, or GMP, for the entire project. For a job this size, the analysis that goes into figuring out the cost of the labor, subcontractors, materials, and so forth is enormous. An official GMP letter from the construction company is also a condition of the bank to close the construction loan.

It's like the old saying: You have to spend money to make money. And boy, were we spending. We hired McHugh, a 125-year-old, family owned and run commercial contractor—one of the largest in the Midwest. We spent six months slugging it out with McHugh and the subcontractors, trying to deliver a construction contract that people believed in, fit the budget, and was in line with our promises to the bank.

While working on the final GMP, it became clear that we were in a really tough construction market. The pandemic had created all sorts of challenges, including disrupted supply chains. That meant we couldn't wait to move ahead with ordering certain parts if we hoped to start work on the tower before 2023—who knew what would be happening in the world then?

We continued to project confidence, a necessary ingredient for any development. But when we still didn't have a signed contract for the construction loan in mid-November—three weeks after it was supposed to close—it was hard not to feel uneasy.

Deutsche Bank's lawyers and underwriting team were not exactly retiring, but my very competent team worked hard to answer all their queries. They fully approved their portion of the loan before Goldman, even though they got in the game much later. The speed of our dealings were interesting considering Deutsche Bank didn't have the best record with real estate loans in the United States. The title of a *New York Times* bestselling book said it all: *Dark Towers: Deutsche Bank, Donald Trump, and an Epic Trail of Destruction* traced the bank's difficult past. Part of my pitch was that the bank would get good publicity for 1000M. Certainly they'd be doing business with somebody more reputable than Donald Trump. Anthony, the Deutsche Bank executive who led the underwriting approval and funding of the bank's share of the 1000M construction loan, moved swiftly and decisively in a very difficult environment to secure the internal approvals to proceed with the deal.

ANTHONY VALVO

Managing Director, New York,
Head of Private Bank - US at Deutsche Bank

This is an extraordinary transaction on many levels. The institution you work with is very important, but the people you access within the institution is equally, if not more, important. Talented people make the elephant dance.

What stuck out most for me in this deal was the vision of the team and their willingness to pull back from the specific path they were going down (a condo development). They had the patience and foresight to change directions one-third of the way into a rental property. It takes a lot of gumption to give back all the deposits; it can also spook lenders.

I'm not willing to take on outsize risk. Banks work on small margins around 2 to 3 percent. If a transaction of this size didn't work, I would have to make $1 trillion in loans to make up for it. So I have to be right.

To get this deal done, we needed people who understood Francis and who could unravel the complexity of 1000M's history. I went through the whole storyline of this property—hearing out the rationality of the budget and the leasing plan, walking the site, meeting with the contractor—in the hopes that what I would find was that the risk was *not* outsized.

Before I went into banking, I was a pilot. I flew corporate jets for about seven years. I still carry a lot of lessons from my former career into my current one. In both you need to be willing to make good decisions with imperfect information; otherwise, you'll never take an airplane up into the air or make a loan.

FRANCIS GREENBURGER: The end of November was quickly approaching, and we were still going back and forth with Goldman. The latest hurdle arose when the investment bank's lawyers wrote up the agreement in a way that was inconsistent with our understanding of the deal. In real estate deals, lawyers retrade everything. They always do, so this was not unexpected. However, the ask was not a small one. If we defaulted on the loan, Goldman wanted the right to take over my $50 million in deposits the bank required I make as part of the deal. There was no way I was going to agree to that, but I proposed a compromise of an offset up to $10 million worth of deposits.

The culture of Goldman Sachs is one in which to work there you need to be the smartest person in the room. That means everyone from the analysts to the bankers to the lawyers are forever trying to prove just how smart they are. The result is that things often become unnecessarily complicated and hard to navigate. Unfortunately, getting the best grades from the most elite educational institutions is not an insurance policy against making mistakes. In 2020, as we were trying to secure financing for the construction of 1000M, really smart people were saying that cities were done. Even in Manhattan, where housing is at a premium, prices for apartments plummeted and vacancy rates soared as residents fled for greener pastures because of the pandemic. "Who would ever want to live in a city again?" financial analysts asked. As it turned out, a lot of people. By summer 2021, New York residential real estate was on fire. Within a span of just a few weeks, the vacancy rates on our rentals went from a high of 25 percent to a low of 3 percent. Apartments, even very expensive ones, were selling right and left. Sometimes you have to question even the "smartest" people.

Goldman accepted my compromise on the amount of insurance in case we defaulted on the loan, and on December 3, 2021, we closed the deal for the $304.5 million construction loan.

Like the end of a jigsaw puzzle, all the pieces started to come together quickly. The city granted us the foundation permit the day before. McHugh were assembling materials and supplies in preparation for going vertical when the crane arrived on site December 13. The money was in the bank, and the market was great. It looked like we had known what we were doing the whole time. Now, we just had to build a skyscraper.

ROB SINGER: The Chicago luxury rental market comeback was remarkable. The 1000M comeback was remarkable. The additional risk we took to redesign the project during the depths of COVID-19 paid off. Having the designs already complete allowed us to react quickly once the rental market took off in 2021. We had our financing and were finally starting construction.

During the pandemic, people talked about the possibility of cities being "dead forever" and so-called urban death spirals. Many asked, "Who would ever want to live in a city again?" Around this time, I read an article that discussed how cities have responded to and recovered from pandemics over the last two thousand years. From the Plague of Cyprian in the third century Roman Empire to the Black Death of the fourteenth century to COVID-19, cities have always faced pandemics and have always rebounded. It might sound ridiculous, but this perspective gave me hope. I thought, at some point, Chicago and other cities would rise again—and we'd be ready for that moment. I could never have guessed how quickly and suddenly "that moment" would arrive.

What's amazed me through the whole process was Francis's trust in me, as well as his mental toughness and flexibility. I've worked with him through both the Great Recession and the COVID-19 crisis. His stoicism is superhuman. He manages an extremely large business with hundreds of assets, businesses, and people. Yet, when these extremely negative market cycles pop-up, he stays calm. He just works through it.

With financing in place for 1000M, our next challenges began to take shape. We now had to build the project—complete it on time and on budget for $435 million. If our plan worked out, and all of our projections came true, we projected that 1000M would be worth at least $100 million more than it cost to build.

Chapter 10

THE REAL WORK, DECEMBER 16, 2021

Construction Begins

FRANCIS GREENBURGER: Today as the crane was installed on site, even I was astonished we had made it to this point. Although the unusual land lease 1031 deal had been my idea, because it had never been done before, there was no reason to think that it would actually work. Yet, here we were: heavy machinery rumbling across South Michigan Avenue; busy construction workers bracing against the wind whipping off the lake; and, behind the scenes, hundreds of others finalizing plans, schedules, permits, and everything else necessary to start construction. The real work was just beginning.

The team at Jahn were working day and night to shift from the design drawings they issued for getting bids on the price of construction to the much more extensive set needed to build the tower. Shop drawing from various vendors were already pouring in. For a project of this size, each subcontractor—from concrete to the curtain wall—prepares their own drawings with even more detail as it relates to their specialty. The architects then return them with more comments. There's a constant back-and-forth over a million details, some of which are mundane, others monumental. There would be upward of five hundred complex drawings by the time 1000M was finished.

The first set, completed months earlier, was essential in setting the guaranteed maximum price or GMP. For projects of this size, construction managers get bids from all the subcontractors and manage the process for an agreed upon price, the GMP. Often, they get a bonus if they bring in a job for less than the budget. If it goes over, they can suffer a penalty. Inevitably, there are change orders issued by the general contractor or construction manager when they feel that the project requires something different than the architect's plans provided for. A change order is a notice to the developer that there is going to be an extra charge because of a last-minute or unanticipated change in the plans. Then, it is up to the developer either to accept the change order and additional charge or to find an alternative approach that keeps the project within its original budget parameters, which is not always possible. The relationship with the developer is more cooperative when the construction manager gets a share of the savings or stands to lose if there are cost overruns.

As one might imagine, arriving at an accurate GMP is extremely complicated and absolutely essential to get right. If you've renovated your own apartment or house, you've probably experienced how some contractors play games. They will quote a price that leaves certain things out, hoping that they'll get the job with a low bid and then they can make money on the other end by stating that the particular addition causing the extra cost "wasn't in the drawings." The same happens with major developments. A cabinet supplier might appear to have a really good product at a good price, except that on closer inspection their bid leaves out the very component necessary for the project. With four thousand units in a building, that would mean a huge cost overrun. That's why it's crucial to have an experienced, hardworking, and trustworthy general contractor, like McHugh, on a skyscraper development.

By all accounts, McHugh is a special company. Founded in 1897, the family-owned, Chicago-based construction-management and structural-engineering firm has been involved in seven of the city's tallest buildings. James P. McHugh started in the family business after fighting in World War II. When he returned from war, he went directly into the engineering program at the University of Michigan where he only lasted two years. He couldn't stand being stuck in school when he knew he was needed at his family's small-scale construction company. With time, he turned the business founded by his grandfather into a juggernaut.

One of the major elements that separates McHugh is that they are not just a general contractor but also a builder. While many GC's do not self perform any work, McHugh can take on the concrete structure work—normally one of biggest components for a skyscraper. That allows them a certain amount of control when it comes to cost and schedule, since they are in charge of one of the project's largest line items. They pride themselves on being able to attract the best subcontractors, who want to work on an efficient, well-run construction site—a reputation that Patty McHugh, who started out her career as a top fashion model, has continued since taking over as chair from her father, who died in 2016.

PATRICIA MCHUGH

Chairman, McHugh

Growing up, I was very close to my father. He would tell me what happened during his day at the dinner table—and he didn't care whether I was four or fourteen, he always asked for and valued my opinion. It helped my self-confidence tremendously in many areas.

I was attending NYU when I was approached by a modeling scout. I started out with Ford, where I was just one of a stable of blonds. That didn't work out, so I went over to Elite where I landed my first paying job for *Vogue*, which was crazy. Still, I ended up at a smaller agency called Pauline's. That woman loved me and pushed for me. It's true

Patricia McHugh

with anything you do—it's important for people to have your back.

I started working quite a bit, but even working and traveling as much as I did, I still excelled in college. My professors at NYU were incredibly supportive of my travels and career. It was the strong, unsolicited advice from my economics professor that I should focus on modeling while I had the opportunity to do so and complete college later that pushed me to model full time while I could. Full-time modeling turned out to be a wonderful adventure—it was the era of the supermodel. While I was definitely a level below the likes of Cindy Crawford and Naomi Campbell, I got to model with them during a glamorous time. And I also learned lifelong lessons on how to deal with many different types of people while still maintaining perspective.

My economics professor gave me excellent advice that led to a wonderful adventure. On the other hand, the advice I received from people in the fashion industry, some very well-known, not to move back to Chicago and work in the construction business was awful advice. "How could you throw your *life* away for a job working in construction in Chicago?" they asked. It makes me laugh to this day. I am sure glad I did not listen to them!

This business was bred into me during all those conversations around the dinner table with my father. Fashion people didn't understand the pleasure of working in an industry that's part of creating a magnificent piece of architecture like 1000M. Everyone at our company and a lot of the subcontractors, we all feel privileged to be a part of constructing 1000M.

FRANCIS GREENBURGER: For its bid, McHugh had upward of twenty-five employees creating hundreds of budgets for this job. They set out to establish the cost based on the block plans from the architect that show how the space will be laid out and basic design concepts. At the helm of this huge undertaking to break down the myriad components of a tall tower into one number to build it was John Sheridan. McHugh's executive vice president was "one of the best conceptual estimators" in the business, or as he called it, "the cocktail napkin number."

JOHN SHERIDAN

Executive Vice President, McHugh

At McHugh we've done a lot of these mega buildings over the years, and we track the unit cost for various pieces of the building. A building like 1000M is extremely complicated, but usually we have done similar components. For those pieces, we have a good idea at a very conceptual level what the cost will be.

For the cocktail-napkin number, we typically break the building into ten different sections in a rigorous process. Those categories are the following:

Site: the infrastructure that connects the building to the city, including roads, sewer connections, and power

Substructure: all the necessary below-ground work, such as deep foundations

Structure: the entire structure of the building

Exterior enclosure: the outside skin or everything that wraps the building

Finishes: everything that fills the inside of the building such as drywall or cabinetry

Equipment: this is equipment incorporated in the construction of the building at 1000M—the building maintenance system, used to clean the exterior is the largest such component

Specialties: this category includes pools and other specialty items

Vertical transportation: elevators, chutes, and anything else needed to move material and people up and down the building

Mechanical, electrical, plumbing, and fire protection (MEPs): various trades

General conditions: management and supervision personnel, cleaning, hauling debris, security, equipment used for construction, utilities during construction, and other miscellaneous costs

At the very earliest stage, when you're being asked what a building is going to cost, you break it down into these categories and track the costs for each based on rudimentary sketches and past experience.

There will usually be a discussion at that cocktail-napkin phase about why the number we have come up with isn't lower. For example, a developer saw we priced fire protection systems at six dollars per square foot. "That's insane," he said. "Our estimate is four dollars per square foot. Why are you so expensive?" It's not that we're expensive. What the developer didn't know was that in all municipalities, fire codes for buildings over five hundred feet tall require a source of water for the fire pumps within three hundred feet from top of the building. The large tanks weren't shown anywhere in the developer's plans at that point in time.

Sometimes the rigorous process we use to identify a realistic number up front hurts us. But I don't want to sit in a room a year later and have people ask me, "Why did this cost more than what you told me?" That cocktail-napkin number needs to be a little conservative. Ideally, as you go through the process of coming to a final GMP, the building becomes less expensive, not more. If we've pushed ourselves to a rock-bottom number, as we develop the drawings with the realities that invariably come to light, we're constantly redoing the pro forma. There's something to be said for setting and chasing an aggressive target. On occasion, when we offer our pro forma, we will say our goal is to take out 10 percent from that number. But the developer must realize that is the starting point and to get the cost down, changes will have to be made.

We put some numbers on 1000M around 2014, but the developer decided to go with one of our competitors. We were upset about it, but the other company had a better number. They marched down the road for about a year more when we got a call to come back in to resume talking about 1000M.

It's interesting how much data we develop from that cocktail napkin. The estimate says what we think is going on. Because the number is broken down in detail, when the question invariably arises why the price per square foot isn't lower, you can have a good conversation about what makes it this expensive.

Once we come on board, the next phase, which we call preconstruction, is much more detailed. Preconstruction is really the process of building the job conceptually. In the first stage, we come up with a cost per square foot based on similar buildings. Then we translate that number into another estimate that is based on the facts specific to the project at hand—like pricing from the world at large—so we can explain to the client the components driving the cost up or down. This is when we start working with subcontractors in earnest.

If you look at the cost of constructing a building, the structure is probably the largest percentage. For 1000M, most of it was concrete with some steel framing at the top and other places. Because we do that work in house, we have good control over the biggest single contract. After the exterior enclosure, the other big contracts are the MEPs. If a tower like 1000M usually costs around $300 per square foot, the plumbing, electric and other MEPs account for $75 per square foot. The MEPs, structure, and exterior enclosure combined account for over 50 percent of the cost of construction.

By the time McHugh signed a GMP for 1000M, those critical trades—the exterior enclosure, the concrete structure, the mechanicals, and maintenance equipment—were locked into contracts. Many of these subcontractors are design builders for their trade and must be brought on board early enough to help develop the permit drawings for the job. For example, the elevator systems are designed by the elevator contractor, not the architect or anyone else. We enter into agreements with the developer so that we can pay each of these trades to start the engineering drawings necessary, not only to determine the price and scope of the work but also to get the permits from the city.

At McHugh, we require subcontractors to go through and initial off every page in the contract specifications, provided by the architect, which applies to them. Over the years, everybody says, "Yeah, yeah, yeah, I got the specifications." But nobody ever reads them. The last thing we want is for the subcontractor to show up on the job and say, "I didn't have that information." We force the subs to mark on each page that they conform or differ with the specs so there's no disagreement once we start building the project.

Hitting the GMP is dependent on good planning—and to the extent the developer then follows the plan. Managing the change process during construction is key. By the time we've signed the GMP, the definition is very clear so it's not difficult to identify changes. If something comes up that's going to cost more, we need to point it out to the developer and explain what changed from the agreed-upon drawings. Then it's the developer's job to make the decision: Is that important enough to spend money, or do I want to do something else?

One of the joys of working with Rob and Francis is that, as much as they're focused on getting the job to the right number, they are not just about pushing the budget. There was a solution for the exterior enclosure that was significantly less expensive than its competitor, however, it was an untested system. Did we want to take that kind of risk on a building of this height where the curtain wall was a signature element? After working through all the scenarios, they went with the pricier system developed by BVGlazing that had been used before on many similar applications. That was the right decision. These mega buildings have a lot of complexity, which Time Equities and JK Equities recognized from the beginning of the job. They didn't get tempted to push that early number down, which often happens, to the detriment of the building's quality or completion.

FRANCIS GREENBURGER: The curtain wall was a never-ending saga. Over several years, McHugh led an exhaustive search for the right exterior facade subcontractor with the help of two of the best exterior facade consultants in the United Staes (Curtainwall Design Consulting and Wiss, Janney, Elstner Associates). The culmination, in 2017, was our decision to go with the subcontractor who came in $10 million less than BVGlazing.

By 2019, however, it became clear that the first contractor had left out a lot of important items in its lowball bid. When comparing apples to apples, the gap in cost narrowed to $4 million from BVG's number. BVG, who we had great confidence in, eventually lowered its bid a little to land on $42 million, and so became 1000M's official curtain wall subcontractor.

In the best of times, choosing the glass for a skyscraper is incredibly complicated. The sourcing is its own adventure. For 50 West, the glass came from Belgium via Spain with certain sections from China. Then there is a rigorous testing process of the product using models. Problems always arise on buildings of this size—particularly when an architect like Helmut pushes the design. Back to 50 West, the thermal load moved the glass curtain wall in unexpected ways. The sun shining on the east elevation of the tall tower in lower Manhattan heats up the glass higher than the north side. The varying thermal loads create movement *within* the system. The result was not structural problems but instead a nuisance noise. We made it clear to McHugh and BVG that we would not accept any snap, crackle, or pop on 1000M.

In January 2020, when 1000M was still condos and COVID-19 was only an abstract idea in the United States, I flew out to Canada with Helmut, Phil, and McHugh's vice president, Randy Bullard, to review a visual mock-up of the exterior facade. Standing in front of the glass and panels of corrugated metal, I was struck by one thought that I immediately shared: "Helmut, this building looks like a radiator." The vertical perforated lines of the spandrels (the metal in between the panes of glass) was like the grill of a radiator. This was not the sleek beautiful monument I hoped to bring to the shores of Lake Michigan. Around BVG's boardroom table, Helmut pulled out his bag of little red pens and started to sketch different ideas for a cleaner design without changing the whole look of the building. Perhaps most impressively, he drew upside down, so that the images were right side up for me, sitting across from him.

PHIL CASTILLO: Helmut believed that you have to be a total architect, which was my training too. You have to do everything. When we were working on the Shanghai New International Exhibition Center, people were amazed when Helmut or I would show up to walk the construction site. But this is so much part of what we do. Good architecture results from attention to detail and to the process. And part of that process is knowing when you need to stand tough and not change the design, or when you should back down and look at it a little bit differently. One of the things I learned from Helmut is that you can always come back with another idea.

NICK ANGELLOTI

Vice President of Business Development, BVGlazing

Phil became one of my favorite people, but the first time I met him I thought he was the meanest guy on earth. I walked into a full boardroom with an engineer by my side and put my briefcase on the desk before introducing myself to everybody around the table. I extended my hand to Phil but instead of shaking it, he went off on me: "That briefcase rolls on the floor! Putting it on the table is a complete lack of respect." He was yelling at me until Randy said, "OK, Phil, leave him alone." I can tell you this: I'll never put a rolling bag on a table again.

We wound up earning a lot of respect for one another, specifically during the eight-month period in 2021 when we had to bring the curtain wall budget down by $10 million without sacrificing the look or quality of the product.

We won the original $42 million contract for the condo building right before COVID-19 put the entire world—including the 1000M project—on hold. When Randy got back in touch several months later, he had good news and bad news. The developers thought they had the funding to move forward, but it was now going to be a rental building. The new budget for the curtain wall was $32 million, but the architect wanted it to look the same. It couldn't look *exactly* the same for that price. Our challenge, however, became the quest to change the design just enough to save $10 million while appeasing an architect willing to make very few concessions.

My parents, Italian immigrants who settled in Toronto, both somehow got into the glass business when they arrived in Canada. My mother was a receptionist at an auto-glass company, and my dad became a site superintendent for a glass contractor. I ended up going to school for electrical engineering, but as soon as I got my first job in the field, I realized it was the most boring thing I'd ever done in my life. I ended up in the glazing trade. Because I had a degree in electronics, I specialized in automatic doors and entranceways, becoming a project manager thirty years ago.

Getting 1000M's curtain wall to $32 million, with the design right, was eight months of constant communication among Randy, Phil, and me. We limited the corrugated metal panels to an elevation up to the twentieth floor instead of the whole tower, which eliminated the radiator look. That recommendation came from Phil, who said it was now a better building because "simpler things are always better."

We also simplified the top of the building. The seventy-second to seventy-fourth floors were extremely detailed and complicated, but I argued that nobody was going to look up and notice anything on the seventy-second floor. By simplifying the top, we were able to save a couple million dollars.

The most significant cost savings came by changing the glass type from a boutique glass fabricator in Beijing known for a fantastic product to an average glass produced domestically out of Michigan. The difference was all in the edge work. In China, they put the sheets through a polishing machine to get a crisp, furniture-quality edge on the glass. If you're within a foot from the glass, you'd appreciate it. But when you're looking at a seventy-two-story tower, the edge becomes an imperceptible detail. Phil wanted it, but in the end, he was able to live without it.

We had to bring our margin down. Our installers, CK2, had to come down quite a bit too. So we each came down 5 percent. We had to make the decision: Are we going to lower our margin to have this job built, or are we going to walk away? One of our competitors had bid such a low number that no matter what I did, we were still at least $2 million more expensive than them. But they had a cheaper product. Francis recognized this and paid the premium to use us.

FRANCIS GREENBURGER: We signed a GMP of $286 million with McHugh for 1000M. By January 2022, the construction company had awarded $34 million for the curtain wall as well as $50 million worth of contracts to the mechanical, electric, fire protection, and plumbing trades. The self-performed concrete structure was approximately another $70 million. While the pandemic meant issues of supply chain disruptions and labor shortages, there was one upside for our project. Because the construction market in Chicago shrunk during COVID-19, McHugh put together a project management team for 1000M that might have normally been on three different projects. They included Project Executive David Steffenhagen, who completed NEMA, a 76-story rental tower kitty-corner to 1000M; Vice President Joel Kuna, who helmed the 101-story St. Regis; and Vice President Randy Bullard, who built the 82-story, Jeanne Gang–designed Aqua Tower. That was a lot of height among those three guys.

Randy and Senior Vice President John Sheridan worked side by side in coming up with the GMP—with Randy often doing most of the heavy lifting. Because he knew all the details of what and why decisions were made, he was the bridge between the preconstruction and construction phases. Joel—who John called "one of the best field-oriented builders that you'll ever come across"—was in charge of everything happening on the ground as director of field operations. "Randy and Joel complement each other well because Randy who knows how to put the job together, the paperwork, the testing, all the things in advance of hitting the job site," John said. "Joel is the one making sure the sequencing is right on the job, that everyone is on schedule, and what to do when we're behind." Together, Joel and Randy were advisors to Dave, who took the lead on 1000M. Dave looked to not just Joel and Randy but to John Sheridan as well for counsel. But at the end of the day, as John said, "Somebody has to make the decision."

RANDY BULLARD

Vice President, McHugh

I grew up in a family of ten brothers and sisters in a rural area of Southern Indiana. After our father was killed in a work-related accident, our mother raised us by herself. We learned at a young age the importance of a strong work ethic and taking responsibility for your actions.

Although I earned my Bachelor of Science in teaching, after college I worked for a large residential builder since I had experience remodeling houses in high school. In my first job, we constructed approximately ninety homes per year until mortgage rates hit all-time highs in the 1980s and the residential market collapsed. I had transitioned into building commercial properties when I met John Sheridan, who convinced me that if I could manage multiple smaller projects, I could learn to build larger projects. My first project with McHugh—where I have been for twenty-five out a forty-five-year career—was construction of the new Chicago Board of Trade facility in Chicago.

When I met Francis, I saw a man with similar traits to Mr. McHugh: a man with true core values, a man with vision, and a man whose word and handshake you can count on.

Francis, Rob, Jordan, and Jerry have a lot of experience, so they knew exactly the high-quality finishes that owners expect to receive in a residential building like 1000M. It was our job to deliver these finishes at a budget we could afford. We proposed ideas for potential cost savings and asked the subcontractors and suppliers to give us ideas too—whether there is a simpler way to construct something or use a more cost-effective material. The question is always: Can we accomplish the same goals in a different way? We put all our suggestions into a master list that we reviewed with the architect, interior designer, and owner. They accepted some of them—such as replacing the folding partition doors in the pool area with two big doors that were more economical. Any ideas that were too dramatic of a cut or change to the look were rejected.

Randy Bullard

Breaking down all the trades, Joel, Dave, and I had to make certain the subcontractors had the right design and were able to meet the right schedule. We also needed to make sure they had the right materials. The supply chain issues during the pandemic that resulted in long lead times for resources made the question of delivery within our time frame even more urgent.

As the materials are fabricated off site, we spend a lot of time in hiring not only outside consultants but with our own quality control teams going out and visiting the factories where the components are made. For each supplier, we need to understand what means of quality control and quality assurance they're providing for: how they source the materials, fabricate them, ship them, and make certain when the assembled product shows up on site, it's what we've ordered; it's what we want to install, and we're not going to have any short-term or long-term issues with it. From the glass of the exterior facade to the flooring type, it's critical to make certain you don't have problems with adhesion, durability, uniformity, and so on. There are so many different things to consider. That's why we put a full-time quality control person on the job to assist our field staff in inspections, field testing, and double-checking subcontractors.

Every project you build is both a lesson on things done right as well as a lesson of what can be done better on the next project.

Chapter 11

"MY BIGGEST POUR SO FAR," JANUARY 22, 2022

1000M Goes Vertical

FRANCIS GREENBURGER: The sun had not yet risen when the trucks began lining up along Michigan Avenue. Four hundred concrete trucks, drums turning in the early morning hours, stretched along the Magnificent Mile, waiting to dump their churning contents into the massive hole in the ground to form 1000M's foundation.

Months of preparation had led up to Saturday, January 22, 2022. It was a monumental task to coordinate with the city before it agreed to close a lane of the busy avenue. There were meetings with the city alderman, the Department of Transportation, the community, and residents of the adjacent buildings to explain in detail the traffic patterns for that day and how we would manage them. Then there was the site itself. Before a single drop of concrete could be poured, caissons were socketed to bedrock, eighty-seven feet below the ground; 685 tons of steel rebar were fitted for the backbone frame; and piping was installed for utilities. Only then were we ready for the mat slab, the twelve-by-one-hundred-foot concrete box that formed the base of the tower.

The mat slab was the task of the day, more specifically the task of Rob Muzenjak, concrete superintendent at 1000M for McHugh Concrete. "A mat slab pour is always a celebratory day because we can see the project really start to come together, and now we can start to go vertical," Rob said. But it was also a stressful pour for him as 1000M was "my biggest pour so far."

For nine hours, Rob led a continuous concrete pour of almost four thousand cubic yards of concrete. Roughly forty-five cement trucks arrived per hour at the site where they were hooked up to four concrete pumps (another pump was parked on the street for backup just in case there was a problem). The suppliers making the mix had to employ facilities across Chicago to provide that much concrete in such a short time frame.

All the hard work done in advance paid off. "The pour went like clockwork," Rob said. "It was almost boring." What was supposed to be a twelve-to-fourteen-hour pour wound up taking just under nine hours. By way of comparison, when McHugh poured 4,600 cubic yards of concrete for the mat slab of the ninety-two-story Trump Tower in 2005, the process took twenty-two hours.

1000M's record-breaking pour was all the more impressive since concrete is one of the toughest trades in construction. A demanding and complex world, it is first and foremost the main portion of the building. That, however, is just the start of its demands. Quality is not easily achieved. If concrete isn't cast and poured properly, once it hardens, the process of grinding, rubbing, and patching the form (or, God forbid, breaking it apart and putting it together again) costs a lot of time and money. "You have one chance to do it right," Robbie said. "And that's it."

ROBBIE MUZENJAK

Superintendent, McHugh Concrete Construction, Inc.

You need to pay attention to every aspect, including keeping the site clean. A clean site is a safe one. I swept the stairs five times that week.

We usually get big water jugs for people on site to fill up their own water bottles, but because we were still worried about COVID-19, we ordered individual plastic ones. Now, the water bottles end up all over the place.

Cleaning doesn't just affect safety. It is also important for the end product. If I place the formwork (the mold in which the concrete will eventually be poured) in a pile of mud, the end product will look like crap. Because we're starting in winter, I also had to make sure there's no snow or ice under the rebar before the concrete arrived, since that could compromise the structural integrity.

I held a cold-weather concrete class with one of our in-house structural engineers to go over temperature's effects on concrete consolidation with my guys. A concrete vibrator shakes the mixture after the pour to move around the cream and consolidate the concrete. The purpose is to get rid of any air bubbles or pockets that compromise the structural integrity and smooth finish when the hardened concrete is stripped. There is a method on how far to take the machine down, how long to agitate to get the mix right. Cold weather means different procedures for the vibrator as well as additives. I am always reminding my guys about concrete consolidation.

It's not always easy to get people who are shoveling concrete for a living all day to listen to somebody about the minutiae. For example, ironworkers like to write directions—like what kind of bar is going to be installed in that spot—on the formwork. But if you write all over the formwork with a paint stick, it will bleed through to the concrete. Yes, we could doctor it later—grind and patch it. Sure. But I don't want that. When people walk through the site, I want them to think the raw concrete could be the finished product, even though it isn't. Although it will eventually be covered, I want you to be impressed when you see the work.

Born and raised in Chicago, I began as a carpenter apprentice, framing houses. I went on to become a tech engineer, drawing all the

Left: Concrete trucks lined up for the mat slab pour

3-D modeling for jobs and then began assisting the McHugh superintendent working on concrete projects. I'm a little bit embarrassed to say I never got a college degree, but I was always educating myself. I took classes for AutoCAD and all the 3-D modeling, as well as construction management classes. A lot of it, however, was learning on the job. I never want to get complacent. Even now, as a superintendent, I want to be the best in concrete.

That's why I tell the foreman that the ironworkers can't be writing on the formwork. "Give them tags," I say. Sometimes I get a look. But my job is to get them to buy into it and make sure the message gets to all seventy-five workers, including the lowest ones on the job who are right out of high school or college, to make sure nobody does it.

A big issue in construction is when foreman don't relay the information to everyone below them. That's why, if there is something critical, we all meet—the subcontractors and all their workers, literally everyone. All it takes is one person who doesn't know how to do something that costs a lot of money.

FRANCIS GREENBURGER: Robbie and the concrete team had a lot more to occupy them than paint on the concrete. The architecture of 1000M—with its negative slope on the south elevation and stepping out of each floor in the tower—made the inherent challenges of erecting any skyscraper even greater. To build this unusual structure, the concrete slabs had to be thicker as the distance between the columns grew on the ascending floors. Those calculations were the purview of the structural engineer, which is basically in charge of designing the frame that holds the building up.

Left: Concrete from Ozinga is poured from the mixer truck into a large pump hopper, which pushes the concrete up through pipes along the building's facade to reach the upper floors

DAVID FIELDS

Senior Principal and Residential Design Leader,
Magnusson Klemencic Associates

Each level, from floor twenty-two through the roof, grows by a few inches compared to the floor below it. That's Helmut vision. When it's built and clad, people are going to find it beautiful. To facilitate this growth, structurally, the columns on the outside of the building also step outward by roughly three inches at every floor, which is pretty unconventional. One result is the span inside the building (the distance between the columns) grows at every floor by the same amount. A few inches isn't a lot, but with over fifty floors it becomes a big difference. The higher in the building, as these spans get much longer, the floor slabs are thicker to bridge the greater distance.

Structural engineering is the application of tangible physics that we all interact with daily, writ large. It's understanding forces and how physical materials respond to equilibrium. It's also translating that knowledge into constructable forms and navigating how architectural spaces need to function. The architect is, without question, the master planner for the whole building. But it takes a team from many disciplines who contribute to make a successful building. Particularly for a very tall building like 1000M, the structural engineer makes decisions and creates forms that also inform the shape of the building and the layout inside of the building. Other systems, like plumbing and electrical, can often conform to what the architect wants. With tall buildings, however, there are limits dictated by physics. Frankly, this building was one of those instances where we had to impose such structural limits that ultimately informed what you see when you observe the building at present. On the south face of the building, starting at level eleven, the building flares out and upward, growing to the south for ten stories before straightening to vertical. The limits of material strengths dictated how steep we could make that slope.

David Fields

Almost all my design work is in high-rise residential buildings. I feel so fortunate to have this job. I decided to be a structural engineer when I was a nineteen-year-old freshman at UC San Diego. I had a summer job as a laborer in my hometown of Juneau, Alaska. While I was digging mud from a sinking foundation, I overheard local engineers talking about how they were going to reframe and hold the building up—and that charted my course. I returned to school in the fall and changed my major from computer science to structural engineering. I realized I didn't want to create something transient, like software. I wanted to be part of creating something permanent, like buildings my grandchildren would see.

I specialize in a mix of Midwest extreme wind design and West Coast extreme seismic design. Wind is the biggest issue with a tall building in Chicago like 1000M. Our early work focused on the portion of the structure and frame that resists wind loads to keep it stable. Following building code requirements, 1000M was designed for the worst storm expected to happen once every seven hundred years. There are national scientific bodies, which study global-storm wind speeds and frequency—and provide this data. The United States Geological Society defines the worst earthquake expected to happen roughly once every 2,500 years. Other groups define the forces, and then we designed the building to withstand those forces.

We also had to design a structure that doesn't make people motion sick due to its wind-driven sway. Even if a building sways a small amount—far below its breaking point—people could get nauseous, so we further stiffened the building to limit this effect. Also at the roof of the building, we designed two tuned liquid sloshing damper tanks. These concrete boxes are forty-five feet long, twelve feet tall, and eight feet wide, and are filled about halfway with water. The length of the tank and the depth of the water are carefully selected (tuned) to match the sway frequency of the building. As the building sways back and forth, the waves inside the tank move back and forth in the same rhythm but in the opposite direction. It knocks the building out of rhythm and reduces motion that can make those in the building feel sick. This is a simple, passive, and highly reliable means of damping a building. There is no mechanical nor powered component that could go out of service. There are very few buildings in the world that have damper tanks. However, my previous project, the St. Regis tower in Chicago, which is 50 percent taller than 1000M, has six of them.

All super-tall towers generally need some kind of damping. There are a number of methods. A classic early example is in the Taipei 101. The tuned-mass damper in the Taiwanese capital's 1,667-foot tower is a 660-metric-ton steel sphere that hangs from 138-foot cables between the eighty-seventh and ninety-second floors. The ball, which is eighteen-feet in diameter, swings in the opposite direction of the sway of the building.

We pick the right method for the space and budget we have. Since McHugh is an expert in concrete, and there is a ready supply of water, the tuned liquid damper tanks made the most sense. Also, the roof provided a linear and horizontal area for dampers. Years before the building is built, we model the structure in our computers to predict how it sways. Once it's built and just before we put water in the tanks, we instrument the building with accelerometers to measure exactly how it sways in the real world. What we're precisely capturing is the time it takes for the building to rock back and forth, called the building "period." For 1000M it takes about six seconds. That tells us how deep the water needs to be in order to be tuned to the building exactly. Stated differently, there's a relationship of wa-

Interior of sloshing damper tanks

ter depth and tank length to how long it takes the waves to go from one end of the tank to the other. The contractor turns on the water source, and it takes days for the water to reach the correct depth in the tank. It's quite maintenance free and low-cost compared to some other solutions.

I get a rush when the structure tops out, and everyone who collaborated on the building celebrates. I get to work with the best architects and contractors in the world. I'm headquartered in Seattle, but I work in every part of the country. Every time I fly into a major city, I keep the shade of the plane window open to look down on the buildings I helped create. Maybe it's a little narcissistic, but it's my ultimate satisfaction. When I was that nineteen-year-old under a building with a shovel, this is exactly what I dreamed of doing.

FRANCIS GREENBURGER: A week after the mat slab pour, they continued excavating around the side of the building to prepare for the foundation walls. The mat slab connected the caissons for the core of the tower. However, grade beams, or reinforced concrete beams, were still needed around the perimeter to tie into the caissons that supported the rest of the building. At the same time, they started the formwork for the elevator core and the structural columns that descend from the fourth floor and split into a striking form, an inverted V, in the twenty-seven-foot-tall lobby with revolving doors onto South Michigan Avenue. To avoid a curb cut that would mar the entrance, cars would enter through a porte cochere behind the lobby.

The foundation work continued for about a month. While the concrete team assembled the core and began on the architectural walls, subcontractors continued to install electric, gas, and plumbing underground, and others readied the framing to go vertical. The coordination among and within the trades was constant. "It's all about communication and planning and scheduling and blood pressure medicine," Robbie said. He called for a meeting with his ironworker foreman and the rebar supplier to find out where they were headed in two weeks' time, so that he could make sure the shop plans were approved and the material was on site when they needed it. "I'm just a small piece compared to what is coming behind me," he said. "I have to get up and out of the way."

"The faster he moves, the more room he gives me to work," said Ricardo Silva, General Superintendent at McHugh, on the 1000M job since day one, before the fence had even gone around the site. "The key right now is to get the concrete done, and then we can put our hands on it."

A native of Mexico, Ricardo left his grandparents' farm where he grew up when he was eighteen years old to find his fortune in the United States. He arrived in Chicago and immediately started working in construction. In 1999, he joined McHugh where Robbie worked as his right-hand man for a few years—including one memorable New Year's Eve. "My wife was waiting for me to come home to go out for a nice dinner," Robbie remembered, "and we were cleaning up a unit from a toilet that exploded."

"You get a lot of respect from all the subcontractors and workers on the field if you never ask somebody to do something you aren't doing yourself," Ricardo added.

Quality and speed were not the only concerns that faced the concrete team. While everybody needed to move in a timely manner and buy into the same best practices, safety was also of paramount importance. Everybody on a skyscraper's construction site is fundamentally a risk manager. Whether it's leaving a piece of plywood unsecured at great heights that could fly off the deck and kill a pedestrian below, or letting trash pile up on the site that causes a fire or tripping hazard—each worker in the field needs to be a safety director and take responsibility for any potential hazards they see.

However, for the concrete subcontractor, safety takes on enormous proportions as the job of framing the walls moves up to higher stories of the tower. As soon as they arrived at the twenty-first floor, they installed eighteen-foot-high windscreens engineered to withstand huge amounts of force. The diaper nuts at the very bottom were made to hold upward of one thousand pounds per thirty feet. The twenty-five-foot-long screens were accompanied by large nets of small mesh to catch any kind of debris that might try to escape. The screens themselves were laced with four-by-fours, so that no wind passed though. Although there are windscreens that allow for air movement, Robbie is not a fan. "The whole purpose is to block the wind," he said.

It keeps your materials, like plywood, from blowing around before the windows and curtain wall are installed. And when the guys are framing the wall, they feel like they're inside the first floor. Because you can't see outside and don't have the wind whipping around, you could be on the ground or fifty-stories up. It can get a little warm in the summer, but it's much more comfortable when it's cold—and we have two Chicago winters to get through.

5807

If you build a car or piece of furniture in a factory, you know what the conditions are going to be every day. The machinery, material, everything is the same. Nothing is repetitive when it comes to building a tower because the conditions are *always* changing. During the first six months into 1000M's construction, perhaps the biggest unforeseen challenge was weather.

By mid-May, there had already been eighteen days of rain. About a half-year ahead of the normal amount of rain days, the level of precipitation was very unusual. "I think we had two days that were observed at full sunlight," Randy said, which he had never experienced during spring weather in his forty years in construction. The issue wasn't the workers, who did their jobs in much harsher conditions. No, the problem was that when it rains, you can't pour concrete. A little drizzle is fine, but a downpour shuts down the site. Rain will ruin any exposed concrete. Twenty-five rain days had been built into the schedule. But with only a week of them left, Robbie was nervous: "Hopefully we'll be able to catch up."

Adding weatherman to his list of tasks, Robbie kept a close eye on the radar, checking for updates hourly. If there was a 30 to 40 percent chance of a shower for an hour, he might decide to go for it. Eighty percent chance of all-day rain was a different story.

Despite mother nature's lack of cooperation, they had still made a lot of headway. They had set a twenty-four-thousand pound triple-separator catch basin that separates the grease from all the kitchens' sinks before the water flows out to the sewer system. They also completed what Robbie called "the best-looking columns on Michigan Avenue."

The "A Columns" were Phil's resolution to a problem that vexed Rob earlier in the design process. Originally, the lobby had one structural column in the middle, which as Rob put it, "messed everything up." The central column meant there had to be two sets of doors instead of a single revolving entrance. The result was hectic and messy. "There had

Left: Ironworker Diana Zastawny

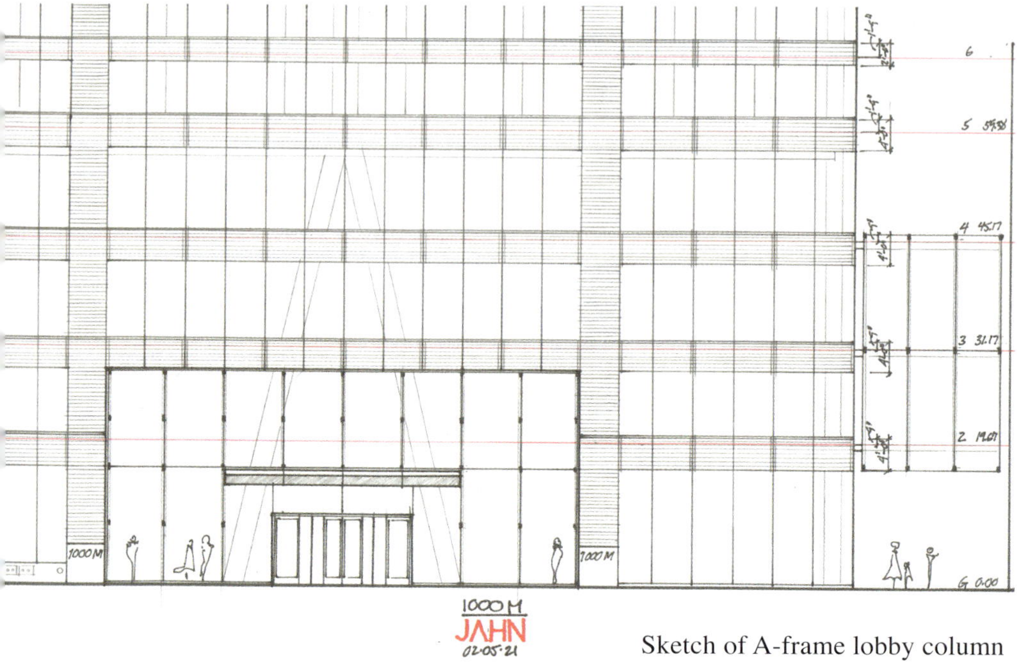

Sketch of A-frame lobby column

Actual A-frame lobby column constructed

to be a better way," said Phil, who did what he has always done when challenged to come up with solutions: sketch.

"Nobody makes sketches anymore," he said. "Everyone gets so caught up in their machines. I may be old, but I think there is a spontaneity and creativity that is lost. Helmut used to say, 'I only draw what I want to see.'"

Phil, who drew two diagonal lines in place of the vertical ones, had an aha moment. He went into Helmut's office with his sketch, and said, "This is what we have to do."

To arrive at a single revolving entrance, Phil split the vertical column into an A-frame that starts in the lobby as two columns on a diagonal and merges on the second floor. From level three, it is one column that goes all the way up to the top of the building. "It will be a great street experience in the lobby," Phil said. "It's a grand space."

But only if Robbie could pour them. Getting perfect consolidation in columns up thirty feet and on an angle was a challenge. Worried that the concrete vibrator might not be effective if used on the entire length of the column, he decided to frame out the entire columns but only pour the lower half first, going down about fifteen feet, then following it up with a second pour for the upper half.

With each completed floor framed, the concrete needed to be pumped higher to reach the next. The spinning trucks on the ground delivered the concrete in a steady stream to a giant stationary pump at the base of the site that forced it up through a pipe, called the placing boom, that ran up the wall. The 18,000-pound placing boom, controlled by remote on the ground, not only made placing the concrete easier but also saved time. Using a concrete crane bucket required time for the crane to go up the tower and down. A crane bucket could only hold three yards of concrete at a time, whereas the placing boom pumped about a hundred yards or eleven trucks worth of concrete per hour.

However, to keep the concrete flowing at that rate, the mixture needed adjusting depending on the height it traveled. It was affect-

Carpenters working to build the next formwork deck. Sheets of plywood and engineered beams create a temporary wooden floor strong enough to hold rebar for the next level's concrete slab pour.

ed by gravity and the pressure of the pump forcing it up through the hose. The concrete had to be modifying to balance out the thickness so nothing got stuck.

A third-party inspector double-checked the concrete at regular intervals for the compression strength. The slump tests, common on any concrete job, measure the consistency of the mixture before the concrete sets. It's a simple process that involves a slump cone, a metal mold open on both ends. The cone is filled with concrete in three stages with each layer compressed twenty-five times after which the mold is lifted off. The way the concrete slumps dictates the viability of the product.

Any high-rise over sixty stories provides unique challenges with establishing control points—a surveying term for the precise horizontal and vertical location of a physical feature, such as the corners of the building. Control points are critical in properly aligning the architectural plans to the real world.

Brian Liebendorfer

BRIAN LIEBENDORFER

Tech Engineer, McHugh

As the tech engineer—or detailer, as we call it—I take established points of the four corners from a grid system created by a licensed surveyor and from there, shoot a precision plumb laser up five floors. Although the laser is supposed to be able to shoot up to one hundred feet with one-eighth of an inch accuracy, we keep it within about five floors, or a hair over fifty feet. After that, I set up the total station—a modern surveying instrument that uses electronics to calculate angles and distances—and shoot the floor sites. The total station is anchored to the core while I walk around with a handheld control for the laser gun. There is a little screen on the handheld device that shows me exactly where I'm going. Every floor, we need to load new data points into the gun. We're measuring a 3-D model with a center point and radius lines that meet the standard points. Every one of those points is its own shot. When you feel good that you've located the points on the deck, you mark them with a pencil. This is to establish the framework for the concrete. Engineers, like me, lay out the plan on the deck for the carpenters, who follow behind to frame it. After the carpenters come, the ironworkers install steel and post-tensioned cables. These high-strength steel wires wound together inside a plastic duct run the length of the building and stay in the framework while the concrete is poured to stay in the center of the slab after its hardened. Once the concrete has gained strength to 50 percent of design strength, the cables are stressed. This helps support the slab in the middle from the columns' supports. When it's fully tensioned, it applies 33,000 lb. of load. The control points on the floor are very important because if they're off, your whole building's going to be off.

FRANCIS GREENBURGER: After all the spring rain, I was just happy to see 1000M was a little bigger than the last time I visited. The facade work couldn't come soon enough. As soon as they closed the building with the curtain wall—floor by floor, several stories below concrete, which continued framing and pouring higher floors—the other trades could work inside every day, rain or shine.

In mid-May, Randy and David traveled to Canada to inspect the glass, and in June the first delivery of 1000M's floor-to-ceiling windows arrived in Chicago. We were lucky to have any glass, which like many other materials necessary for construction was in short supply during the pandemic. If you look at towers going up, you will often see a lot of concrete and no glass during construction. That has to do with a delay in the glass, which often comes from overseas. Because BVG was one of the biggest customers of the company manufacturing the glass for 1000M, they were able to get the order, all in one run, ahead of time, and store it in their factory. "We didn't have to worry about the supply chain," Nick Angelloti explained.

A curtain wall, however, isn't just glass. In the case of 1000M, aluminum was another important component—and one that skyrocketed in price during the pandemic. In what Nick called "mere luck," they locked in the rate when BVG was awarded the contract, right before the price rose about fifty cents more per pound. "That was the last project we were able to do that," he said. "Even now, if I order material for a project, whatever the value is on that date that it's run is what they charge us. We lucked out when it came to being able to lock in the tonnage because there was quite a bit of aluminum on that project."

On the 8th floor, crew members remove the formwork from the freshly poured west wall behind the indoor lap pool. The form held the concrete in place during curing, shaping this structural wall.

A subcontractor can have everything ready to begin their part of the project, but if the sequencing is delayed, they can't start anyway. Even with the heavy rain, McHugh Concrete was on schedule, which meant that the curtain wall installation could begin on time.

BVG's highly engineered unitized curtain wall system made installation relatively straightforward. Fifty-five drafts people and engineers worked months drawing every single element from the aluminum extrusions to the silicone gasketing. Each component had its own unique page from which the Canadian company then manufactured them. In the factory, they assembled, crated, and wrapped panels that "essentially snap together like LEGO," Nick said. A five-by-twelve-foot plate is dropped into place, hooked onto the building's floor slabs, and hangs there until more panels are put in place on either side of it. The process moves clockwise around the building, one floor at a time. Once a floor is fully snapped together, the work jumps above it and, again, moves clockwise.

No matter how straightforward a process may be, there is always the possibility of something going wrong. When 50 West was completed, the engineer we hired to test the curtain wall refused to sign off on the project because the sealant was discolored. He suggested the sealant, mixed in China, was not properly formulated, and so its strength might be compromised. The whole building was already built!

We immediately sent a sample of the discolored seal to Dow, the chemical company that manufactured the product, and asked them to test it. Their analysis was that it was still at least four times stronger than what was required by the New York City buildings code. The engineer wasn't satisfied. He wanted us to "clip" the building. To put metal clips around each window would not only ruin the look of 50 West, it would also cost more than $5 million. We all went nuts, including Helmut.

Rob found a leading national curtain wall engineering specialist, who repeated the tests and concluded that 50 West's exterior passed. We still had to persuade the original expert, since the buildings depart-

Lobby windows

ment didn't allow you to change engineers in the process. We set up a meeting with the national curtain wall expert and at least two dozen architects, engineers, and other professionals with the original engineer, who, ultimately, was unable to defend his position. That didn't mean he relinquished his power. He ordered tests on every floor, so we sent men crawling all over the outside of the sixty-four-floor building. We also had to remove a section of the curtain wall, take it off site, and test it. After many months and much expense, the original engineer finally gave the sign-off.

Politics finds its way into everything, even curtain walls. BVG understands this. That's why it doesn't handle the installation of the curtain wall system it manufactures. Instead, they subcontract that part of the job to a strong local company. "Through the years, I've learned if a bunch of Canadians fly into a new town to install curtain wall," Nick said, "it doesn't work out."

STEVE TOBOLIC

General Foreman, CK2 Contracting

1000M's unitized curtain wall system is unlike the window-wall systems we typically do on other jobs. In a window-wall system, a cell receptor and head receptor are installed before the glass. These receptors run continuously along the perimeter of the building. On each floor the window is broken up by the concrete slab, which we cover with a slab cover. A unitized curtain wall system is all glass, from floor to floor. The window is held in place by hooks that attach to preinstalled clips on the floor. Made out of aluminum, they connect to studs preinstalled in the slab, which stick out of the concrete. A washer and nut tighten and hold the clip, which is then responsible for holding the window in place.

For the installer, a window wall system is much more labor intense than a unitized one. On the other hand, the unitized system is a more expensive facade because of the design and the fact that there is more glass. If I were to build a skyscraper, I would go with the unitized system. It's cleaner because it comes together with fewer parts. Everything in a window wall system has to line up. With the system we were using on 1000M, it's pretty much straight up the side of the building.

Before anything, we worked on a layout to locate everything we would need while installing 1000M's curtain wall. The layout establishes how things are going to run up the building continuously, so

Right: CK2 crews work in sync across multiple floors, carefully lowering glass panels and securing them into place. The teamwork between levels ensures each window is safely guided and mounted into the building's facade.

you want to dial that in as perfectly as you can. We began with the prep work for the level-one windows, getting all our sill starters in and making critical seals. Once we got that first stack going, it was basically off to the races. My crew on 1000M averaged between twelve and sixteen guys on any given day. They were eight floors behind the concrete pour. So as fast as they could pour, we could install.

There's a lot of quality control involved in the process. The window comes from the factory, which has its own quality control. But when it arrives at the site, we do our own inspection of the window panel prior to setting it. We're checking the appearance to make sure there is no damage. Then we need to make sure the layout of the windows is spot on and that they are engaged properly.

When installing windows, it really comes down to the seal, to make sure no air or water will penetrate the system. That's number one. An independent company comes to the site to test the windows. They build a chamber on the interior side of a window covered in plastic. They spray water from the outside while at the same time vacuuming air from the inside to pull moisture. If there's one spot of water, it's a failure. In our contracts, it's written that testing occurs every four floors at random locations. So far, we haven't failed a single test.

I've been doing this for twenty-five years and have worked on plenty of skyscrapers. My father was in Ironworkers 63, the union of architectural ornamental ironworkers—that's how I became one. This is different than structural or welding and burning ironworkers. We are highly trained to install metal windows, curtain and window walls systems, and metal stairways, gratings, doors, railings, fencing, elevators fronts, and building entrances. When you've been doing this job for as long as I have, you think you know what to expect. But there is always something new. Maybe it's a new product that you haven't worked with before or something of that nature. What I appreciate the most when I get to each new job are the views. They never get old.

Workers from CK2 installing windows

FRANCIS GREENBURGER: In summer 2022, a symphony of pings, dings, thumps, and thwacks accompanied me on my tours of 1000M. The soundtrack was heartening. So was the view. Construction had passed the top of the rectangular podium base and was moving on to the tower. I needed all the good news on the construction side that I could get.

On July 27, the Federal Reserve hiked interest rates by three-quarters of a percent point for the second consecutive month. The fourth rate increase in five months was the Federal Reserve's attempt to douse rising inflation. Consumer prices jumped 9.1 percent in the previous twelve months, marking the highest inflation rate in more than forty years. Our construction loan was a floating-rate loan, so rising short-

Concrete leveling, 2023

term rates affected us—as they did anyone borrowing money, whether that be a mortgage or credit card debt.

We bought a rate cap, or interest-rate insurance, when we closed the construction loan to protect us during the time of construction. We wouldn't even begin to draw down the loan until October since the terms dictated that we use equity first. Still, as rates went up that meant each day of work cost us more. When we were $100 million into our construction loan, if it were at a rate of 5 percent, that would work out to $15,000 a day in interest. In other words, if there was a delay of two weeks—due to weather, supply chain issues, a mistake, or countless other reasons—that would cost us $200,000. Those numbers only get worse as time goes on.

The real exposure to rising interest rates, however, would be *after* 1000M was built, when looking for a mortgage. When we first began working on the financing, we assumed that rates would be higher in the future; but how much higher? Rates were certainly going to continue

Dave Steffenhagen

to rise as the Federal Reserve battled stubborn inflation. The first two quarters of 2022 were negative, the technical definition of a recession. Nobody knows where rates will be two years in the future, but if the recession lasted for another few years, the Federal Reserve would likely lower interest rates. With a little bit of luck, that might coincide with our locking in a fixed loan for 1000M.

Developers spend years designing, planning, and financing, and conceptualizing the market. In the end, however, the tradespeople are the ones who turn an idea into reality. I was humbled to watch them work. Whether it was the window installers hanging off the ledge hundreds of feet up in the air or electricians snaking miles of wire through carefully planned tunnels in the concrete, they are the ones who coordinate their expertise on these hugely complicated front lines. It's their hard work that turns an idea into a reality.

A building grows in Chicago

Chapter 12

CLOSET CRISIS, MAY 2022

If It's Not Yet Built, It's Not a Mistake

FRANCIS GREENBURGER: One day you're looking at an architectural plan, a piece of writing, or an investment, and it looks fine. Then you look at it the next day, and you think, "What the hell are we doing?" The change in perspective can be attributed to time, a good night's sleep, or new information. The more material and detail you're dealing with, however, the harder it is not to lose perspective. Imagine, then, looking at hundreds of pages of complicated floor plans for a skyscraper—and deciding to rip them all up.

That's exactly what happened in late May 2022 when Rob Singer was beginning to work on the renderings for 1000M. Creating the digital images that appear as if they are real estate photos of an existing construction is a whole process that will be treated later in this book, but on this particular day, Rob was checking out what the rendering views were going to look like within a unit when he noticed something odd. Reviewing unit 6904, a one bedroom on the sixty-ninth floor, he was pleased by the kitchen, hall closet, pantry, nice living room, bathroom with two sinks, big shower, and linen closet. But there was only one closet.

"Wait, what is this? This can't be," he thought, wildly moving the cursor around and refocusing his eyes, scanning in vain for the other missing closet.

Rob's panic wasn't a deep-seated passion for fashion or home organization. When he began the redesign of 1000M from condos to rentals in fall 2020, he worked closely with Mark Ziemke and Aaron Galvin from the Chicago real estate company Luxury Living who made the case that we could outperform the market if we focused on built-in desks and, even more importantly, closets. Part of the argument came from brokers reporting back on the reactions of prospective renters visiting NEMA. At the luxury high-rise, they would be given a lengthy tour of the many, many very impressive amenities. After an hour of being dazzled by the coworking space's nine conference rooms, sports bar with old-school arcade games, a chef's kitchen cooking demonstration space, and much, *much* more, they finally arrived at the units only to find the kind of closets you typically find in any rental—and not many of them. Facing the prospect of one reach-in closet with a single rod, the potential renter walked right out. No indoor lap pool could make up for that. Brokers told us over and over that the developers couldn't get the pricing they wanted because people couldn't actually imagine living there.

Over the year it took to redesign 1000M, the team worked hard on the sizing and mix of the units with a strong focus on closet space and built-in desks. In this luxury high-rise rental, every bedroom would have two closets at minimum. Except unit 6904 only had one.

Rob assumed this was just one mistake. After all, there were 738 units, including almost a hundred unique unit types. He immediately called Phil: "There's one unit that's screwed up."

Of course, it wasn't just one unit. They quickly realized the problem was much bigger than that. As they started to review the plans, they found many units with only one closet. Remedying the situation was going to require serious surgery. They would have to move bedroom walls, which then would mean new furniture layouts. Suddenly, the plans they thought were done completely unraveled.

This wasn't just an issue of more work for the architects and designers. Although, there was that too. We had already signed contracts with many of the subcontractors and were almost five months into construction. It wasn't easy for Rob to come to me with $2 to $3 million in unbudgeted costs to optimize the closet space throughout 1000M. The big picture is that we convinced lenders that our units were impeccable, and we had hired people to build them, only to kill the whole plan. Rob could have let it be; but he believed passionately this would be a major misstep. I gave him my blessing, and he returned to the drawing board. Literally.

It wasn't as if Rob didn't already have enough on his plate. "We have fifty line items, and every one is a battle," he lamented on the design process. There were countless details to consider and negotiate among Jahn's architects, Kara Mann's team, and the subcontractors. It was one thing to design a space but a whole other thing to get it into the plans. The design drawings had to be sent to the trades not only for pricing but also for shop drawings showing how they were going to build the design. Hundreds of hours of scoping and drawings going back and forth ensued.

Here's just one example out of countless: Kara Mann's office specified a one-inch side panel for the side frame that boxed in the kitchens in all the apartments. However, the contractor explained the wood we had chosen only came in three-fourths of an inch, or 1.5-inch widths. A heated debate flared up instantly. There was talk of custom building the frame at an additional cost of $400,000. "Stop it!" Rob said. "We can't get derailed. We are building a skyscraper."

That is the constant pressure of the developer pushing the team to come up with ideas, solve the problems that crop up, and move the project forward. The process on settling on the final kitchen design—one that would be repeated 738 times throughout the building—was a prime example of the tension between paying attention to the details without getting drowned by them.

Only about a week before Rob made his closet discovery, we did a site visit to look over the kitchen mock-ups, real-life kitchens built by different contractors vying for the job of making all the kitchens for 1000M. There were several reasons for the mock-up process. The main purpose was to review the materials—such as the countertops, cabinet laminates, and flooring—which always look different from their samples when they are used in actual construction. You can also tweak unintended mistakes that always arise from even the best of plans. A contractor can get the dimensions of the drawings on paper exactly right, but in real life everything is a little different. It's better to work out any minor details *before* a thousand cabinets arrive on site. Lastly, giving three different contractors the same exact information for each to build their own mock-up creates a sense of competition among the contractors that is beneficial to the whole process. Building their kitchens in the same space made it nearly impossible for them not to keep tabs on one another's progress.

The mock-ups were housed in the office building we owned adjacent to 1000M and the genesis of the entire skyscraper enterprise. The three different kitchens, plopped in different areas of the raw space, were out of place, like sets built for a TV sitcom or play. A group of fifteen people—including Rob, Mindy, Phil, Randy, and me—gathered around one of the kitchen islands. Unsurprisingly, there was no shortage of opinions.

"The outlet needs to be a 3 gang and run vertical, not separate," Phil said, looking at the switch plate one of the contractors had installed on the backsplash. "I don't know how they screw these things up, but they do."

That was right before he pointed out that the drawer pulls, which were currently in metal, should be the same wood laminate as the cabinet and drawer faces to maintain a minimal look. There were argu-

Francis and Philip discuss finishes

ments about where to put the seam in the countertop and where the joint should go.

Then we debated the laminate patterns. Amberleaf's Palomino was a good color but too knotty. Kara Mann liked Lineadecor's Nuovo Gray Oak, wire brushed for what her people called more of a "realistic wood look." The treatment they thought made the laminate appear more natural to Rob and me just looked like scratches on the cabinets.

"I find it disturbing," Phil said.

We vetoed the scratched-up Lineadecor laminate.

Next up were the countertops, always controversial in any kitchen project. The options were two engineered stone products, Caesarstone and Silestone. Once 1000M went from condos to rentals, we had lost most of the heavily veined marble that is part of Kara Mann's trademark look because it was no longer cost-effective. Instead, we hoped to mimic the look of real marble with engineered quartz. When we saw the two products, there was no question to me that the Caesarstone option was the winner. However, it was also $300,000 extra for all units. I'm a huge believer in budgets, but you also need to keep an eye on where it makes sense to splurge. These countertops were considerably better looking and would make a big impression to would-be renters. In the scheme of 1000M's total budget, $300,000 was a drop in the bucket. (Of course, too many drops and you have a flood.) Plus, we were saving $100,000 on our flooring choice.

I asked the group what they thought, and we were all in consensus: everyone liked the Caesarstone better.

"Done!" I said, reveling in a rare moment of unanimity.

"Now it's time for me to negotiate," said Quintin Huckabee, the McHugh project manager in charge of the subcontractors working on 1000M's interiors.

QUINTIN HUCKABEE

Project Manager, McHugh

On a job this size, there are typically a few project managers who have a batch of trades that they manage. I specialize in the interiors, which means I manage the drywall framing, painting, flooring, and millwork. I personally had to review the drawings. Then I went out for a request for proposal (RFP) on each of those trades. I sent that out to multiple qualified bidders for each of the five trades I had to award. On average there were five qualified subcontractors for each specialty, which gave me a total of twenty-five subcontractors I was in communication with until we started getting companies under contract.

Obviously, everything has to start with a budget. I get handed the budget when I get handed the drawings. During the bidding phase, I start to receive numbers from subcontractors after they've reviewed the drawings. I review those numbers to make sure they are within a certain range. We typically like to see the range within 10 percent of

Quintin Huckabee

our budget, because that tells us everything is apples to apples. Once I get numbers that come in close, I set up scope reviews. Out of all the companies, I typically pick three to four to sit-down with and discuss the ins and outs of the job. I make sure that the company is qualified, and that the team they are bringing out to the site is qualified and a good fit for the project. A job like 1000M is going to run for multiple years, so you want to interview the people you will be working with. We run the process just like you would a company.

After the interviews, I send out the final request for best and final numbers. That's where everybody sharpens their pencils. Are there any savings that can give a subcontractor an advantage as far as pricing or schedule? That's the last step prior to awarding the contracts. We incorporate all of the information into an award recommendation, which goes to ownership for review. I've yet to have an award recommendation get kicked back to me from ownership because we do a very good job of vetting the potential subcontractors. We tighten up the details to make sure ownership and the architect get what they want, and everything fits within the budget.

For 1000M, we awarded the millwork contract last. The millwork is all the custom woodwork within the building. The lobby desk Is a staple millwork piece in this type of building. That was a very detailed portion of the 1000M job because there were very extensive amenity levels in the building, and those are all millwork. For me, the millwork was the final piece that tied everything together.

My job is to make sure everybody's happy. You have the architect who wants the best of the best because their name is on this project. They want the product to be a staple for their firm. Ownership wants a really good product as well, but they are also driven by the dollars. We had to go back to the drawing board on some of the selections because the job was bid over a year earlier. Once we got to final contract time, we ran up against a lot of escalation costs because of the current climate. The cost of goods spiked because of inflation, COVID-19, and the war in Ukraine, which drove up gas prices. When we originally budgeted, we had a pretty good number. But once we went out to the final bid, a year later, those costs went up about 10 percent. If the budget I was managing was around $50 to $60 million, 10 percent was a lot of money.

There was a lot of back-and-forth on what the finishes of the unit case work for the kitchens should be before we got everyone to a happy medium. In the mock-ups, we asked to see a couple different finishes. Nothing was operable, but the point was to see what the rooms would look like. One was the high-end finish. Another was a value-engineer finish, meaning you "engineer the value" of a particular piece to try to bring it within budget. When ownership looked at the countertops mocked up, the overall decision was to go with what they picked from last year, regardless of price. They wanted to make sure the finishes were well thought out throughout, even if there was an additional cost based on escalation. It came down to the veining, which is a pretty big deal. Just one or two details can expand a lot throughout this type of building.

I'm used to working out details and understand that they won't be done in the first pass. For me, the toughest part of the job is telling the contractors who didn't get the job. All the companies spend a lot of time and effort trying to be the one chosen. They study the drawings, get numbers from suppliers, and build mock-ups in order to give an accurate bid. It's a lot of man-hours, and many of these companies are family-owned businesses for whom a contract like 1000M can be a big deal.

I try to let them know how much we appreciate the work they've put into the process and that although it didn't work out on this one, hopefully we can work together in the future. I have an extensive sports background, so I understand that even if you put the work in, sometimes it doesn't pan out. You don't win every single game you play. You don't hit every single pitch you swing at. But keep going. Eventually, the time will come, as long as you keep at it. But you definitely see the light in people's eyes go out when they see they are not going to get to work on a seventy-three-story skyscraper in Chicago right on Michigan Avenue. That's a marquee job for anybody.

I'm from the South Side of Chicago. When I was a young kid, downtown was an amazing place. Downtown is a different world from the South Side. The opportunities are different. So it's amazing for me to work on a project like this and to be a part of reshaping the Chicago skyline just fifteen minutes from where I grew up.

I got into this line of work because of my father. He was an engineer, who worked on the upkeep of buildings' mechanical, electrical, and plumbing systems. He knew what I was good at and steered me in the direction of engineering, which I had an interest in because I grew up around it. But he directed me into this field and encouraged me to hone my skills.

I bring him out to the site whenever I can. He loves it. Having been around a while, he puts his two cents in and challenges me—in a good way. He'll ask me a question to see if I'm on my toes. We have that good back-and-forth. I know he's proud, but he just says, "You listened." That's good enough for me. He's done a lot for my brother and me, so it's good to see him happy.

FRANCIS GREENBURGER: The trades with signed contracts to work on 1000M were getting nervous in fall 2022 when Rob was still not done with his "process of unit optimization," the redesign of 738 apartments to include more and better closets than originally back in May. "When are you going to be done?" they asked over and over. "We need to know what we're building, and we need to get prices."

Fair enough, but there were so many units to review. To make matters more difficult, 1000M wasn't a standard building where the apartments stack up uniformly. Because of the way it stepped out and twisted, there were all kinds of configurations that Rob and the team had to go back through. To make matters even worse, every time the architects sent him updates, he would kick them back—they just weren't good enough. The team pushed back, but Rob pushed harder. They shifted, pulled, robbed, and rearranged. They added sliders and scrutinized every inch of the building. He drove them and himself crazy. Meanwhile, time and money were adding up.

In addition to offering excellent amenities, our plan for 1000M was to win on unit design with the best layouts and finishes in the market. When it came to unit layouts, we focused on optimizing the normal areas such as kitchens, bathrooms, living rooms and bedrooms. But we also zoned in on maximizing and optimizing the closets in each unit—particularly in our studios and one bedrooms, unit types that often don't get reasonable closet space. We wanted several closets in each unit, and we wanted them to be large closets. To accomplish this, we had to squeeze down other areas in the units, which was a trade-off we felt made sense. Closets are among the very highest priority for renters, and they are often overlooked by builders and designers.

Typically, builders and designers include one or two small closets in a unit and then finish them with one wire shelf at the top and a hanging bar. For 1000M, in addition to increasing the number and size of the closets in each unit, we wanted to deliver high-quality, built-in closet organizing systems. The built-in organizing systems were a real differentiator from other rental buildings. The shelving, shoe racks, hanging areas of different length, and other elements meant our tenants would not need to use traditional dressers that take up precious space.

We contracted with Modular Closets, a New Jersey–based company that makes predesigned and custom-made closet systems. An entire second design process began, looking at each individual closet to see how they should be built out. The result was a 615-page document that Rob combed through as thoroughly as he had the unit layout. Rob and Moe Feder, Modular Closets "chief closet officer," went over thousands and thousands of closets, one by one. Rob rejected many of the initial designs. The shelving wasn't tall enough, or it was too wide. There were no shoe racks. The look wasn't right. These were relatively large open spaces—some of them with windows—that deserved as much consideration as any other part of the apartment.

"There's no point in making a big closet and just putting in a few wire shelves and a rod and telling people to piss off," Rob said. "We're here, we're committed—let's go all the way."

MOE FEDER

Chief Closet Officer, Modular Closets

I found out about 1000M in a roundabout way. I had met Jordan Karlik's brother, who suggested I reach out to him about a development project on Long Island. As soon as I googled Jordan, I said to myself, "Wait a second. He has a monster going on in Chicago." When I realized Time Equities was the lead, I remembered we had done a couple small units for them in New York City a few years ago. I took that as an opportunity to send an email to Jordan and Rob Singer, pitching Modular Closets to work on 1000M.

We used to partner with a factory in China, which made the closet parts that we warehoused and shipped and installed nationwide. Because of the uncertainty of importing from overseas and the cost of tariffs, in 2019 we moved all our production from overseas to Lakewood, New Jersey. Everyone thought we were crazy, but we took a risk, and thankfully we did. Making everything in house, we dodged a major bullet during the pandemic. Shipping containers cost $25,000 for other closet companies that were also quoting a five-month lead time, when ours was a month.

During COVID-19, so many people lost or left their jobs. Only about one employee out of the two hundred who work at our company left. I think part of that is because we run the business like a big family, including serving hot, catered lunches every day to every employee. We realized that the small things you do for your employees can have a big impact. If you're working on the production floor, you're not making a million bucks. Saving ten dollars on a burger means something.

Rob and Jordan were looking for a product for their closets similar to what we were manufacturing, a semi-custom prefabricated closet that was somewhere in between cheap wire shelving and a $10,000 custom closet. When I showed them some ideas, they didn't

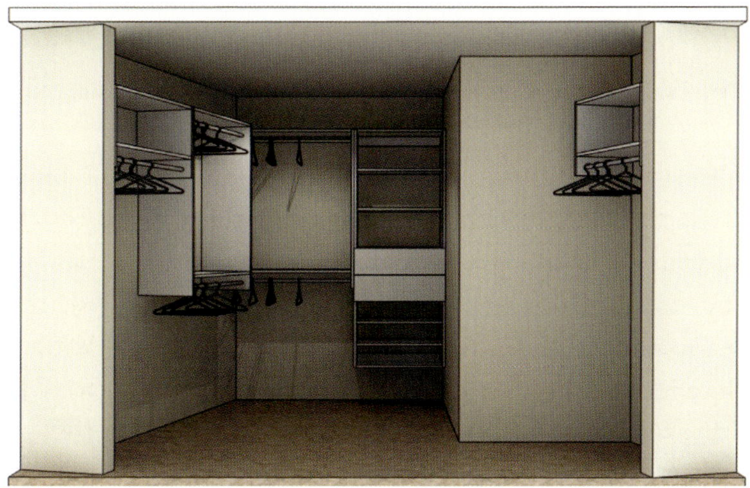

Modular closet plan

mince words: "We need the next level." Modular Closets is not customized. To keep it simple, we're a prefabricated system that comes in three colors. But for a whale like 1000M, we could make an exception. I'm not a design guy; I'm just a sales guy, but I suggested we try to match the closets to the same finish as the kitchen cabinets—and they loved the idea. We typically don't do that, but we changed our entire run to match the finish. For months and months, I went back and forth between Rob and our factory getting the finishes to line up.

So many people think, "It's just a closet." But a lot of thought goes into building a good closet. What happens if the architect squeezes an eighteen-inch closet in a corner. That looks great on paper, but if you work with closets, you know that when you put a hanger on a rod with clothing, it's going to stick out another seven inches, and you'll have tenants complaining that they can't close their closet doors. We caught that in a bunch of units and made changes. In the end, we probably had north of six hundred pages of shop designs, because we wanted to make sure every single one was right.

A good closet is one that maximizes space. Today, we live in the online shopping age. I think I paid the fuel bill for Jeff Bezos's private jet last week with the number of packages that came to my front door. The question on everyone's mind, including tenants, is where do you put everything?

For developers, if they want to maximize closet space, they need to think about it during the construction process and not after the building is done. That's when you can figure out, "Hey, if we add another foot to this wall and make the door swing go out instead of in, we can add another piece here." It's those small details that make all the difference.

FRANCIS GREENBURGER: We spent countless hours picking out furniture, finishes, and choosing every detail and design element to make the units so chic and polished, renters would be wowed. But the closets, desks, and other built-ins also served a different function. The better the use of space, the happier people could be in smaller units than they might have thought.

In a rental scenario, the whole idea is to get prospective renters into the smallest amount of space they can live with. That might seem counterintuitive. One might imagine brokers trying to upsell clients from studios to one bedrooms, or from one bedrooms to two bedrooms, and so on. But that's not how it typically works. Most people are driven by a monthly budget. If you can show them a smaller apartment in their price point that works for them and they fall in love with it, that is the ideal situation. Efficiency, functionality, and taste are directly related to maximizing the rent per square foot.

1000M's studios provide the closets of a one or two bedroom; the one bedrooms have the closets of a two bedroom. The customer pays less than they planned by going down in unit size, but the rent is still a premium over what a similar unit would have garnered without the considered layouts and closets. That's why you need to get people into the space. Because when someone in Chicago sees a $2,700 studio on paper, their reaction might be, "What? I'm not paying $2,700 for a studio!" Once they walk into this studio with its sleek millwork, views of the lake, and unheard-of closet space, they hopefully get it. In order to get this style, including closets, desks, and everything, they might have to go to a one or two bedroom somewhere else—and spend $3,500 or $4,000. When they compare this against a regular studio, it's like they are getting a luxury one bedroom at a discount. We're giving them the same utility and a higher design in a smaller space.

Nonetheless, someone might love the studio but can't abide having their bed in the living room. If they want to know what else is available in the building, the broker doesn't have to immediately jump to a one bedroom. We have a lot of steps before we get there. Again, the point is to get them into the smallest space in which they can be happy. There's a junior one bedroom where the bedroom isn't separated from the living space with a door but instead, millwork shelving. It functions as a distinct bedroom and has a great flow, but it's less expensive than a one bedroom. The client might have seen five one bedrooms in the same price range, but they are excited to take the junior one bedroom that has less actual square footage because it is so much nicer.

"The style is great; the views are great; the amenities are insane. On top of all that, the layouts work, and in particular, the closets will allow us to generate a premium," Rob said after working around the clock for months, moving walls, adding closets, shifting this and that, until finally finishing the redesign in October 2022.

It remained to be seen. The market could very well disagree and show that people don't value closets as much as Rob and the guys at Luxury Living believed. The financial success of a real estate deal is incumbent on making a lot of good decisions. In a tower like 1000M, there are fifty thousand decisions to make—and even if you make 90 percent of them right, the deal can still wind up a failure. You need to be close to 100 percent right, and then these deals can be *really* successful. No one knew that better than Rob, who waded through many potential mistakes every day and pushed aggressively until they were rectified. On the issue of the closets, however, he was convinced. "I've been in this business for more than twenty years, and I've never seen something like this," he said. "This is revolutionary."

The trades move up with the tower

Chapter 13

THE TRADES, OCTOBER 2022

The Podium Completed, Work Begins On the Main Tower

FRANCIS GREENBURGER: The work on 1000M had been very steady. With the concrete team diligently leading the way, the curtain wall installers had been moving in lockstep four floors below. (While pouring the newest slab, the construction workers were reshoring the cured concrete four floors down before moving up.) Everything was fitting together just as it should, so that by early October the top of the core reached the twenty-fourth floor, with crews pouring the twentieth-floor deck and building the formwork on the twenty-first. The twenty-second level marked the beginning of the main tower and a faster building schedule. With fewer walls and columns to worry about, they accelerated to a three-day pour cycle. While concrete continued to flow from above and glass started to envelope lower down, the other trades could begin to move up the building. The order of the work was electric, plumbing, HVAC, framing, drywall, taping and skim coating, tile, flooring, carpentry trim and cabinets, appliance installation, painting, and finally cleaning. Each trade tread the same ground but forged wholly different paths.

Rob Singer with Rebecca Paley

The journey of 1000M's plumbing and electric began way earlier than the tower's rise, when water, sewer, and power needed to be brought in from the street to the site. Site utility subcontractors negotiated streets and alleyways—not to mention municipal authorities—to hook up the connections to the utilities essential to the building and the construction process.

The plumbing subcontractor on a skyscraper does so much more than install toilets or hook up washer-dryer units (even if there are hundreds of them). For 1000M, AMS Mechanical Systems was responsible for all the plumbing-related items within the property line—including piping for the swimming pools and installing stormwater storage tanks underneath the building. As a "design-build" sub, they were in charge not just of installation but also design of the plumbing systems.

Over the past twenty years of high-rise residential construction, a few different methods have evolved. In "typical" construction, the scope of work for the electrical, plumbing, HVAC, fire protection, and other trades are designed and coordinated by engineers hired by the owner and architect. These engineers of record submit their calculations and drawings to the building department and are legally responsible for the designs—including providing special professional liability insurance just in case there are issues with their respective designs. In a design-build process, the selected subcontractors are assigned the responsibilities as the engineer of record.

Hiring design-build subcontractors is not just about transferring the risk from the architect and owner to the individual trades. There are other reasons to allow subcontractors to complete the design of their component of the project. The first is timing. It often takes longer for the owner to hire outside engineers to coordinate with the architect, complete their design, and submit their drawings to the building department for permit review (it's a process that takes several months at minimum). Then it takes another few months for subcontractors to provide pricing for those drawings, which is necessary for establishing a realistic budget to secure financing for the project.

Historically, subcontractor design-build is more cost-effective and quicker. However, every decision comes with potential advantages and drawbacks. Some developers feel more comfortable hiring outside engineers as another layer of expertise and protection. As a check valve for potential risks with a design-build sub, the developer hires a third-party engineer to review and monitor the subcontractor's design as well as selection of equipment, the installation of work, and commissioning of each of the systems. On 1000M, we used design-build subcontractors for the curtain wall, building maintenance equipment, elevators, fire protection, HVAC, electric, and plumbing.

FRANK KUTROMBIS

Plumbing superintendent, AMS Mechanical

I was born in Athens, Greece, and moved to Chicago with my family when I was eight years old. I became a plumber after working as a laborer doing water and sewer work for the city, which laid us off in the wintertime. My buddy suggested, "Why don't you come and be a plumber instead?" I said, "Sure, I'll try it," and I've been a plumber ever since. That's twenty-nine years.

No plumbing job is alike. Each has a different layout and different piping systems. AMS Mechanical is a huge company that does plumbing, mechanical, HVAC, electrical, fire systems, even nuclear. I've done everything from high-end residential jobs to food processing facilities, hospitals, schools, you name it. We are a complete piping shop that handles not just water but waste, venting, and other systems.

We had one in-house engineer and two CAD designers working full-time on the engineering drawings and modeling for 1000M's plumbing during the design-build process. Since we were the plumbing subcontractor when the building was condos, we had to redesign the whole plan for rentals, which had more apartments but fewer bathrooms. Our main challenge wasn't so much about changing the plans—instead, it was getting all the piping, plumbing parts, and fixtures.

Throughout COVID-19, we struggled with supply chain issues, because, while a lot of piping is made in the United States, specialty fixtures and many other necessary items for our trade come from overseas. Everything is preordered months prior to the start of construction to make sure that we have all the material ready to install. The minute there is a redesign, we have to break out all the parts and reorder again because they aren't the same. It throws everything off. We had to change how we did business because of the pandemic. Normally we try to stick to one or two suppliers, but we probably reached out to a dozen suppliers in Chicago to find who had stock and who didn't. It was really a huge struggle.

Another struggle was labor. Once construction resumed, Chicago became so busy. It was hard to get labor on projects because everybody was swamped. We needed thirty field employees, a full-time

Alex Flores, a plumber with AMS, sealing cast iron drain pipes

project manager, two designers, and one engineer to work on 1000M. As a subcontractor it's important to work as team with a great general contractor, like McHugh, to get through such a monumental job.

I love being a plumber. You walk around a building, and everybody thinks, "Oh, you're a plumber. You just change toilets." It's so much more than that. You have to know how to do math, and how have to value engineer systems to work within budgets. It's a great trade. I haven't been laid off in my whole life. A general contractor we used to work with—an old man, who was a great guy—he told me, "There are two things you need in life: a great doctor and a great plumber."

FRANCIS GREENBURGER: Everyone might need a great plumber in their lives. Without electricity, however, you won't have a building in which to put the pipes. Before the project can start, the electric subcontractor must create a whole plan for the temporary power needed during the construction process. They begin where the skyscraper does, underground, running lines that energize the project. In the case of 1000M, they brought three thousand amps from ComEd to power up the crane, the man hoist, temporary lighting and outlets, and all the different equipment required throughout the project. While running the temporary power, Gurtz Electric, the company in charge of all the electricity for 1000M, also worked on the building's permanent electrical needs from fancy light fixtures in the communal library to computer lines for individual apartments.

By November, they were finishing up the vault to house the permanent power source from ComEd, which they established only a few months later. It was a major milestone to get the permanent electrical service established that early in the building. Many new developments in Chicago had to wait until the very bitter end, right before they were ready to be occupied. Part of the problem was a shortage of chips and switch gear for the electrical panels. Then it takes almost three to four months to switch over from temporary power to permanent power. Get-

Danielle Villareal of Livewire Electric installs cables

ting the elevators and mechanical systems set up on permanent power earlier in the process would save a lot of time toward the end of the project and allow us to turn over the 1000M quicker.

TIM MOYLAN

Design Development, Director, Gurtz Electric Co.

Installing electrical systems in a skyscraper of this scale presented several challenges, particularly when integrating advanced technologies to allow residents to control their entire living environment directly from their smartphones.

Acting as both the electrical engineer and contractor, Gurtz Electric completed a full 3-D model of all the building's electrical systems to ensure they were accurately integrated with the complex architectural elements. The unique design of 1000M, which cantilevers over the adjacent building and rotates as it rises, added significant challenges to the installation process. The rotation of the tower required careful coordination to ensure that the electrical components aligned properly across the entire structure, especially as the building twisted upwards. Routing the electrical systems through the building's geometry that shifted with each floor required expert planning and highly skilled techniques. We also employed advanced tools such as reel in reel cable-pulling systems and drones for site assessments.

There were also significant difficulties with getting material and equipment due to supply chain issues. Switchgear and meter banks lead times grew to well over one year from date of order. By coordinating with the project team and utility early we were able to get all large electrical equipment, designed and released for production to ensure no delays in the project schedule. Pre-fabrication was another crucial strategy in ensuring that installation of the electrical systems didn't disrupt the overall construction schedule. Building them off

Russel Clark of ANH installs heat pump

site, especially the large conduit systems, reduced on-site labor time and material waste significantly.

1000M's electrical systems—from vehicle charging stations to controlling the lighting or calling the elevator from your smartphone—set a new standard for luxury high-rise living in Chicago.

FRANCIS GREENBURGER: While the electric subcontractor was partially working within the concrete slab, most of the other trades trailed after the pouring process. Three to four weeks after a floor was poured, the HVAC sub started its work installing the ducts and pipes that would eventually run up the entire height of 1000M to bring heating and cooling throughout the building. Most people think of HVAC—which stands for heating, ventilation, and air conditioning—as the systems that control the temperature. But the trade is also responsible for helping control building pressurization, or air flow, which in a tower as tall as 1000M is a big job.

PATRICK BELL

Project Executive, Hill Mechanical Corp.

As the HVAC contractor on 1000M, we were given a narrative put together by one of the owner's outside consultants outlining the basic premise of the design for the HVAC. The standard plan—which outlined the type of equipment for the individual units, as well as the central plants that generate the heat or cooling for the units at the top of the building and in various mechanical spaces throughout the building—is commonplace in the industry now for residential high-rises.

From there, we, as the HVAC design build contractor, determine the required heating and cooling capacity of the building. We work hand in hand with the architect on the characteristics of the facade—or what we would call the skin of the building—and what the internal usage of each space is within the building. This is primarily an all-glass building. The glass has a certain U-value, which is a term for how much heat it allows into or out of the building depending on the time of year. We run a "heating and cooling load calculation" for the entire building, down to the individual room. The summation of that is the total heating and cooling needs for the building. This is the up-front design work that's done well before construction starts.

From there, we approach equipment vendors to find out who can provide what we need. As we start to make agreements with vendors to purchase equipment, we start shifting from the design process to the building process. We work with McHugh's superintendents to have all our material ready to go on site. All the trades collaborate through BIM, which stands for building information modeling. It's a digital process that involves 3-D modeling, creating a construction workflow, monitoring and altering the workflow when construction begins, and providing an important resource for future repairs or renovations. Coordination is crucial for safety and efficiency. BIM coordination keeps field workers on site from clashing by designating the time and space for them to arrive with their equipment and install it.

HVAC systems are, for the most part, vertical. There's very little horizontal piping or duct work on any given floor. We fall into sequence with concrete when enough floors have been poured so that we can begin installing the vertical ducts and piping. While we're installing ducts on one floor, we're staging our material for the next. It was a very repetitive process, because most of the floors were very similar to each other. We got into a good rhythm.

Hill Mechanical, which is based right by O'Hare Airport, celebrated its eighty-second year in business in 2023. In the sixteen years I have been with the company, HVAC systems have become much more efficient. Older residential buildings typically operated in heating *or* cooling mode, and that was dictated mostly by the time of year. Individual residents didn't have the option to decide the mode in their units. Now the standard is a heat-pump design, which allows each individual user to operate their equipment in either heating or cooling mode any day of the year, no matter the weather. Another change from when I first started is Wi-Fi-compatible thermostats, another new standard that allow residents to adjust the thermostat for their individual units remotely.

For 1000M, we took advantage of another innovation in HVAC equipment by using a low-temp heating hot-water system in which one pair of pipes handles the heating and cooling at any time of year. This is different from running two separate pipes with one designated for cooling and the other for heating. The drawback to this system is that it requires larger components in some of the equipment. But the benefit is a reduced footprint when it comes to the piping. Running one pipe instead of two adds up to significant savings in a seventy-four-story building and easily offsets the cost of the large equipment. That wouldn't necessarily make sense in a ten-story building.

A low-temp system has its place, and 1000M is a place where it was a good fit.

There is always a challenge as the HVAC contractor putting the Swiss Watch together. At the top of the building, everything ends in the mechanical room. All the vents and ducts, electrical conduits, and plumbing pipes: They come together and must be joined into common systems. The mechanicals and utilities also fill the ceiling space of each floor. We work with the owner and architect to give them the desired look, ceiling height, and open space—while still finding space for everything that needs to fit for the building to function.

Another important aspect in a tall building is trying to combat the building pressure so that, among other things, you don't have a huge rush of air into the building at the lobby. Paths that go all the way from the bottom to the top of the building are the biggest sources of air movement. In a high-rise those are the elevator shafts, which basically act as chimneys if the pathway isn't blocked. We worked with the architect to seal off where the air could travel from the elevator to other spaces. They added a second set of doors to the lobby, so that the elevator lobby is isolated from the main lobby.

My favorite part of the job is tracking the progress. When the building starts to get buttoned up, and we turn our systems on and control what it feels like in the entire building, that's the payoff at the end. But I get the most satisfaction watching the building rise vertically and knowing our systems are going in just as quickly.

FRANCIS GREENBURGER: Witnessing a tower's rise, the satisfaction of seeing a job from start to finish, particularly a building grand enough to transform the skyline, is the best part of the job for many who work on skyscrapers. Rachael Dolecki is no different. She loves to watch the projects she visits grow to completion. But in many ways, her job *is* different. As McHugh's safety manager, she is there to be "a fresh set of eyes." As tradespeople repeat the same work day in day out, the

Crane lifts building materials

routine can cause them to miss potential hazards. Rachael gently—and not so gently—coaxes them out of complacency "to make sure that everyone's going home back to their families the same way they came on the job first thing in the morning."

RACHAEL DOLECKI

Safety Manager, McHugh

Honestly, I got into the trade as a joke. About fifteen years ago, I said to my father, who was a laborer for thirty-eight years, "Hey, Dad, why don't you get me a job flagging?" I had been at home, raising my three daughters. Two days later he called and asked me, "Are you serious?"

"Are *you* serious?" I said. "Are you going to get me a job?"

"You can start tomorrow."

"OK, let's do this."

That's how I got my foot in the door. I started off flagging at the Ann & Robert H. Lurie Children's Hospital of Chicago for their concrete subcontractor. At some point, my father asked a safety director he had known for years to keep me in mind if he needed anyone. It just so happened that one of the safety manager's jobs was going through a transition where that safety manager was moving to another project. They trained me to replace her, and I wound up becoming a permanent part of that crew.

Out of my eleven years that I've been in safety with McHugh, eight of them were solely with our concrete division. I was a safety manager, overseeing the crew on one job site from the start to finish of the concrete work for that project. Over the last two years, my role has changed. Now I'm on our general contractor side, so currently I oversee ten jobs, including all the subcontractors working on those jobs. I'll pop into a project, walk it from top to bottom, making sure that everyone's following the proper safety protocols, and comply with OSHA.

McHugh sent me to OSHA 145—a 145-hour construction site safety supervisor program. For six months, I attended the program twice a week. I would work all day and then go to school for three hours. As a safety manager, I do a lot of continuing education to stay up-to-date and fresh. I'm not the only one who has to keep learning. McHugh just sent all our general contractors, superintendents, project managers, project engineers, and assistant project managers to redo their OSHA 30-hour training—which I did as well. The OSHA 30 certification—which includes general worksite safety practices, workers' rights, and employer responsibilities—does not technically expire. But standards change, so it's always a great refresher.

When my girls, who are now teenagers, were younger, and people asked them what their mom did for work, they would say, "She babysits grown men." It's literally what I do.

On a site, if I come across anything that is a safety issue, I address it with the individual right then and there. A lot of times, I'll go to their foreman or their supervisor to let them know what I saw as well. But my job often begins before the subcontractors have even mobilized out on a site. I sit in on preconstruction meetings to let everyone know what the expectations are in terms of safety and what they need to turn in to us prior to mobilizing the site. As the work progresses, I conduct safety inspections, where after I walk a project I report everything I saw from good to bad—and how we are going to correct any issues.

I handle any safety concerns or injury incidences that occur out in the field, but more than that I work hard to prevent them by reminding the workers to follow best practices. Probably my most common reminder is for people to put their safety glasses on. I give props to the window installers. They're on that leading edge. Again, when people are looking at a building going up, those are the first people they see. They are typically careful to tie themselves off, but I always remind them to ask themselves, "Are you tied off properly?"

It's very rare that I would be disrespected, or the guys want to be jerks to me. I have a very strong personality, and most of these people know I'm going to give you respect as long as you respect me back.

Housekeeping is always a top priority on any project. When it comes to high-rises, there are special safety concerns. It's imperative that they have a windscreen or perimeter protection that encloses the top of the building. You must be really careful not to lose material off of the building, especially when you're at such high heights. 1000M is also in a very high-profile location, right on Michigan Avenue, where during summertime there are crowds out and about at different concerts and festivals. There's always something going on within the area in which the building resides.

A lot more eyes are on you when you're doing high-rises. People are curious and want to know what's going on, which can lead to a relatively new safety issue: drones. When we were working on another luxury high-rise, we had a lot of drones, operated by kids or residents in nearby buildings, flying by the tower crane moving with material. Drones have become such a big issue that we had to include

Abel Lozano on top deck securing cables to pour concrete columns

it in our orientation. We tell the workers that if they see a drone, don't do anything. Call us on the radio, and we will find out who is flying it. We had to do that a number of times.

I have a very good working relationship with all our different subcontractors to the point where they feel comfortable enough calling me if there's something happening or they have questions. Ultimately, everyone's goal is to make sure that we get the work done, get the building up, and get it up in a safe manner.

FRANCIS GREENBURGER: When 1000M is finished, it will be almost entirely windows. All that glass needs to be cleaned and, at times, repaired. For this type of structure, a system much more complicated than a window washer is required to maintain the building's enclosure. Tall towers—such as 1000M, Aqua Tower, or the St. Regis—all have maintenance units that are a permanent piece of the building. They are essentially mini cranes installed on the roof that fold out over the edge of the building to allow a skilled worker to reach every corner of the facade from inside a rig. There are many factors that go into the engineering of these units, which are also used to replace any broken glass. The weight of the glass is used to calculate the capacity of the boom. Some systems drop straight down. But architecture with positive or negative sloping exterior walls are especially complicated for this already complex equipment. McHugh brought the building maintenance unit subcontractor into the process early on, because it recognized a machine that feeds from the top of the building and moves in as it descended was no small feat of engineering. Martin Wunder, technical director of the company that built 1000M's maintenance unit, said that while the inward slope made the system "very complex," it is "nothing out of the norm."

MARTIN WUNDER

Partner, Global BMU

After the 2000s, the skyscraper industry exploded. There was a big race in the skyscraper industry to produce the highest and nicest buildings in the world. It wasn't just about building towers two thousand feet high, the geometry also became really crazy due to new materials and architects designing in 3D with computers. I have done over one hundred building maintenance units in hot spots all over the world, like Shanghai, Dubai, New York—wherever skyscrapers are built. Curved, sloped, bent, nothing is straight anymore. Forty years ago, every building was square. These days, they are pyramids, twists, or spirals.

Every building over seventy-five feet needs a concept to clean the windows. Access to the facade is mainly done by building maintenance units (otherwise known as BMU). They are typically placed on the roof of a skyscraper or super-tall building and from there people can access the facade by a platform that is suspended from a rig connected to the roof with wire ropes. From there people can inspect, do minor work, wash the facade, or even replace the glass panes. The biggest challenges for those of us who make these units are the new height increases and the geometry of the facade.

For 1000M, we made drawings and began communicating with the structural engineer to figure out how the building could take the load of a 60,000 lb. machine on its roof. We also had conversations with the curtain wall subcontractors since the machine needed to be guided along the facade. Although buildings now come in all shapes, 1000M's slope was unique and so the design of our system had to be as well.

An independent third-party engineer will check the drawings and approve them before they go to manufacturing from equipment suppliers in Europe. These companies in Spain, Germany, and Fin-

Greg Kusior building staircase forms for concrete pour

land manufacture the entire BMU according to our drawings and ship them to our testing facilities in either Barcelona or Madrid. There we assemble the BMU, test it, dismantle it again, and ship it to the project site. 1000M's BMU was manufactured in Barcelona. In the more than two decades I have been working in this field, Spain has become the main country where BMUs are produced. There are a lot of manufacturers around the world; still, most come from Spain. Even the most famous German BMU company was taken over by a Spanish company a few years ago.

1000M's BMU has a double-telescopic jib arm that, when the machine is parked on the roof, telescopes in to make it almost invisible. The machine should not be seen from the street or neighboring buildings when it's not in use. When folded in, the arm has a reach of only thirty feet. When it's working, however, it reaches eighty feet. The jib arm, which telescopes out horizontally from the center of the roof, acts as a guide. It can turn 360 degrees, allowing the system to reach the entire envelope of the building.

On the roof car, there is a drum hoist with four steel-wire ropes that go through the jib arm. The platform connects to the ropes that are suspended from the tip of the jib. The platform moves with the shape of the building by being tied toward the facade as it slopes inward. The people in the platform have a mechanism to tow the platform toward the facade at each level. They plug lanyards from the platform to intermittent stabilization anchors to keep a strong connection to the building. This is especially important in case of wind. When lowered down a few hundred feet from the roof, the platform can easily swing from the wire ropes—and that would be very dangerous.

Every three floors, the people manning the platform tie back to the building, using special lanyard pins recessed into the facade so they are not visible. We build the pins and deliver them to the curtain wall subcontractor who installs them. A minimum of two people

Jeff Nemcek cuts rebar for lower floors

operate the platform, descending or lifting themselves up with the ropes using an electrical signal. Auxiliary ropes attach to the facade.

There are many more safety features in these BMUs than when I first started working in this industry. All the movements of the machine are limited by sensors. Lasers monitor the distance of the platform from the building. If the operator booms out too far or fast, the sensor stops the machine automatically. The day we hand over the machine to the client—the BMU is lifted onto the roof by the tower crane during the construction phase, and our team assembles it—there is extensive training of the safety protocols every user has to follow. And training needs to be renewed frequently.

Just as the height of skyscrapers has grown, so has the size of BMUs increased a lot. The machine is also equipped with an additional glass hoist unit. This enables the owner or facility manager to exchange broken glass panels anywhere on the facade. This hoist can lift 1,000 lb. in addition to the weight of the platform. This is a crucial innovation because once a tower like 1000M is built, you can't reach most of the facade by crane—it is too high. You need this equipment to repair and maintain a building 790 feet up in the air. The size and weight of the latest BMUs has also necessitated other considerations, such as developments in noise and vibration protection for the floors below. These machines are placed on the most valuable part of the skyscraper: the top. People in expensive penthouse apartments are not going to be happy if they can hear or feel a sixty-thousand-pound machine rumbling.

FRANCIS GREENBURGER: BMUs are incredible mechanisms that are vital to the life of a skyscraper. One thing they aren't responsible for, however, is evacuation in case of a fire. 1000M's entire fire-maintenance system was managed by United States Alliance Fire Protection. The largest full-service fire-protection solution provider in the Midwest, USAFP designs and installs fire-sprinkler systems, fire-detection systems, fire-suppression systems, and special-hazards suppression systems for commercial, industrial, and residential projects.

JASON AGOSTINHO
Senior Project Manager,
United States Alliance Fire Protection

We install fire-sprinkler and life-safety systems in everything from single family homes to mid-rise and high-rise buildings. Warehouses, freezer-cooler institutions, schools, hospitals, we do the work in anything that can and does use fire sprinklers. Most municipalities now mandate the use of fire sprinklers, even in single-family homes. Not all, but the smart ones use fire sprinklers in all their institutions because fire sprinklers save the lives of the people in the building by allowing them to get out in time.

On 1000M, a steady crew of five to six handled all the fire-sprinkler installation. Working alongside the plumbers who initially provide the main water service to the building, we put in all the sprinklers and

Jason Agostinho

the miles and miles of sprinkler pipe that service the building. The fire alarm system was done by a separate contractor since that is on the electrical side.

The height of the high-rise will dictate on how many fire pumps are needed in order to get the proper pressure all the way to the top. Sometimes, we need to add water tanks partway up the building as an additional water source. In this case of 1000M, we utilized the slosh tanks at the top of the tower as part of our design to protect the building. The damper tanks, designed by the structural engineer to prevent the swaying effect, provide the water supply for some of the fire pumps. The first twenty-three floors are fed from a fire pump on the second floor, and then from there, the remainder of the building is fed from the two tanks up on the seventy-second floor. This technique has been used before, but 1000M is a unique version. Because of the magnitude of the slosh tanks, which take up a portion of the seventy-second floor, we have two fire pumps up at the top.

Knowing the systems that we install save lives makes you feel like you have really done something.

Chapter 14

ART IS AN AMENITY, 2022

Considering 1000M's Art Program

FRANCIS GREENBURGER: The spring sunshine poured through the glass facade of 1000M's lobby, which was beginning to hint at its future grandeur as a group of us huddled together to discuss the art that might somehow enhance such an imposing space.

"I was looking online at an artist's work this morning. Ron Klein. He's an artist who travels the world collecting pieces of nature and man-made objects, like sea pods or old baseballs, and transforms them into these incredible wall sculptures. I'm thinking that maybe one of them wants to be on that big wall up there," I said. "Something on a giant scale that complements the height experience."

"Are you still interested in doing the marble blossom?" asked Lynda Dossey, a principal at Jahn. One of the major architectural minds behind 1000M, Lynda was referring to a large stone sculpture in my collection that I had previously mentioned.

"Yes. They could go together."

This was no small discussion we were having on site in May 2022. The lobby is people's first experience of a building. Without being overpowering, its art must be visually interesting and have longevity. That is just the aesthetic piece. Then there are all the practical considerations, which Lynda quickly began to elaborate on.

The "marble blossom," two marble flowers weighing upward of 5,000 lb., was so heavy it needed to be integrated into the structure of the building itself. The architects and contractors had to reinforce the floor beneath the sculpture. "We have a design that we can do to translate the load of it down to this matt slab," Lynda said. "We better figure it out and let the guys know. We sent that information over to McHugh in December, but now that we're in the zone we need to bring it up again."

"We might need a crane to get it in here," someone from the construction crew said.

"I think we'll have to wheel it in through the front before the revolving front doors go in and everything is closed up," Lynda said.

"That's not a problem," replied David Steffenhagen, McHugh senior project manager. "What'll be hard is getting it out again. You'll have to take the building apart."

If *The Earth Laughs*—the marble blossoms—worked for 1000M, it would have a long distance to travel from its current location on a Broadway median near New York's Lincoln Center. That was nothing, however, compared to the length of my relationship to the artist.

I first met Jon Isherwood when we were on the board of the Triangle Artists' Workshop, an artist residency program in upstate New York, founded by the famed British sculptor Anthony Caro and the collector Robert Loder. Eventually, I founded my own not-for-profit arts center—Art Omi, a sculpture and architecture park, gallery, and residency programs for international architects, artists, dancers, musicians, translators, and writers, that resides on 120 acres in the Hudson Valley. Jon went on to run his own residency, The Digital Stone Project, that focused on sculptors, but we never lost touch. We became very good friends, and I have collected a lot of his work.

Jon's work has evolved over time. He would often cut the stone in a way that contrasted with its natural shape, making interesting interven-

The Earth Laughs arrives

tions with a chisel or a stone cutter. With the advent of technology in stone cutting, a whole revolution that Jon was a part of, his sculptures took on more elaborate, softer-edged forms. To turn the material into shapes that belie the marble shows both a lot of stone craft as well as a forceful imagination.

His sculptures were featured in one of the first exhibitions for Time Equities Art-in-Buildings program, which brings contemporary art by emerging and midcareer artists to nontraditional exhibition spaces. In this case, his show was held in the lobby of one of our properties, 125 Maiden Lane. I purchased a piece from that exhibit and continued to collect his work, displaying it in my homes.

Perhaps the most meaningful sculpture was the commission to memorialize my late wife, Judy Willows, who passed away from cancer, and our young son, Alexander, who tragically drowned. In a large hollowed-out boulder on Omi's property, visitors can sit and read a tribute to their lives.

Our relationship and conversation about art has lasted more than thirty years, so it wasn't out of the norm when Jon approached me about financing a major piece for a public art installation of his work. The costs of fabrication for pieces of this size are significant. Jon was dreaming of this piece and wanted to get it done. If I would pay for its fabrication and an artist's fee, I would own *The Earth Laughs*. At the time, I wasn't thinking of 1000M; I was just planning to add it to my collection.

JON ISHERWOOD

Artist

What would later become *Broadway Blooms* began with a call from Anne Strauss. A curator who retired from The Metropolitan Museum of Art's Department of Modern and Contemporary Art, she was lending her talents to the Broadway Mall Association, which maintains and beautifies five miles of Broadway's malls, the landscaped islands between roads that stretch from Washington Heights to Manhattan's Upper West Side. Tasked with curating exhibits along the avenue's median, Anne and I had worked together many years ago.

Not long after, I jumped on the subway and got out at 103rd Street and Broadway to meet Anne. It was summer, and the horticulture that the association developed along the malls was very beautiful and inspiring. I was reminded of a commission for the Peninsula Hotel in Beijing where I wrestled with the question: What is this moment where travel ends and rest begins? It's that transition from the road to the hotel lobby to your room. For the Peninsula's front courtyard, I created pathways with floral patterning influenced by the beautiful tranquility of Monet's *Water Lilies* and the lotus ponds at the summer palace in Beijing.

Walking south on Broadway, among blooming flowers in a busy city center, I was also struck by the generosity surrounding me. Planting and tending to a garden that anyone can enjoy is a gift. "Well, what the hell is an artwork?" I wondered. "Isn't it some kind of gift, too? Isn't it meant to give?"

My last revelation came when I landed at the Lincoln Center for the Performing Arts, the association's southern border. I have always loved the moment where people throw flowers on the stage after a performance. The bouquet lands and just sit there, presenting itself in an interesting reclining position.

With all these ideas scribbled in my notebook, I started to think about what I would actually make. As an artist with a history of abstract metaphorical work, I was taking on a new challenge by embracing a more representative idiom. I didn't want to go as far as photorealism or, on the other side, a very specific narrative of the flower, like Georgia O'Keeffe, but rather somewhere in between. I found that inspiration in Édouard Manet's still lifes of flowers, specifically *Branch of White Peonies and Secateurs* (1864). His gesture in the paint, the building up of the surface and texture, has a form that makes the picture come alive.

Once I realized the theme, look, and feel I wanted for this series, I started the process of "sketching" ideas. I use 10 to 20 lb. chunks of water-based clay that I shape and model to figure out what this thing is going to look like. Often I come up with ten or fifteen ideas, I build molds around the clay, empty out the clay, and pour in plaster, which I further refine by filing and sandpapering after the plaster mold has dried. When I feel I'm getting close, I scan the molds with a 3-D scanner. Once the models are on a digital file, I use various 3-D modeling programs to pull and push the form around a little more. Most architects and a lot of artists use these software programs. At a certain point, though, I get a little anxious that I've lost the art to the virtual space because I can't reach into the screen and touch the thing. That's when I create a 3-D print of the piece, not so large, just enough to see it.

It's a beautiful moment when you get it back in your hand again. At that point, your hand becomes the eye. It's a sensor telling the brain if the object feels right. You're looking and feeling at the same time. We do it with a lot of things. We do it with shoes or the vessel we want to have breakfast out of. All those things have some correlation between the hands and the eye.

When I'm feeling pretty good about the direction, I begin to think about scale and size—and then go shopping for marble, which is great fun. *The Earth Laughs* is one part Rosa Portogallo marble and then the other marble is from Italy. I work a lot with stone from Italy, more specifically the Apuan Mountain region above Carrara, Italy, whose marble is a household name.

It's a special and dramatic place. The weather can be stormy and moody in one moment, bright and sunny the next. The geology of the mountains is equally dramatic. Not only are there a lot of quarries, but some have four or five different types of marble, which is quite un-

1000M Lobby, Left: Ron Klein, *Power of the Small*, 2023-2024
Center: Jonathan Isherwood, *The Earth Laughs*, 2020-2021

usual. Calacatta Gold moves to a purer white Carrara or even a green-hued Cipollino Verde Vagli. It's bizarre the amount of different activity in a region made famous by the likes of Bernini and Michelangelo.

Picking a piece of stone from the quarry is also an imaginative process. Looking at the block, you try to conceive of the grain, coloration, and vein inside. Choosing marble for a sculpture is so different than for, say, countertops, where you want the stone to be brilliant and the grain relatively consistent. Instead, I want some of that movement from the earth's formation.

The actual sculpting process begins with an extensive conversation with robot programing technologists at Garfagnana Innovazione computers. We work with software to program seven axis computer-controlled robots. Creating the "tool path" defines the program for the path, speed, depth, and other factors of the stone cutting. When we initially cut the block, I start wide and go narrow. I do that on two other orientations, so I'm basically left with a cone-like shape. At that point, I stop and ask myself how the machine should start to remove the material and what type of milling head I should choose for the robot. At this point, we are still roughing it out, trying to get close to two or three inches from the final form.

After roughing out the form, I decide upon the surfaces. With *The Earth Laughs* it was very important to me to leave some markings inside the flower, starting in a repeated pattern from the very edge of the petal and moving toward the center, to give the petals a feeling of opening and expanding out. For the outer surface, I only wanted to bring out the beauty of the stone.

I go through another milling process within about half an inch of the final surface, and then spend the next six months doing the final shaping—grinding, cutting, carving, sanding, and so forth. I make a fair amount of change, because the object changes in its real scale from the smaller clay models. When confronted with the stone, I really wanted to make the petals round so that they felt like they are generating out in bloom.

When the final polish is done (I tend not to put a high polish on the forms; you don't need to see your own reflection!), I leave the sculpture for a few weeks. There's that moment when you think, "I've seen it and seen it, and I can't see it anymore." You need some time to come back to the work with a fresh eye, look again, and have fresh revelations. With *The Earth Laughs*, the petals' edge needed to curl back a bit more.

It's a beautiful process of sculpting. Sometimes people, used to the traditional craft of hand and tool, are critical of my use of technology in assisting the stone carving. "You didn't make it, a robot did," the argument goes. Apart from the fact that I created the design and wrote all the tool paths, I consider the technology itself an exciting part of the art. When one looks back through the traditions of sculpture, at Michelangelo's work or the marble hands series by Louise Bourgeois, you see the tool marks, the way the chisel hit the stone. Milling with a robot allows for incredible, unprecedented precision, because it won't make a single mark other than what you've programmed. Unlike when a traditional craftsperson carves from a model and there's often a little bit of natural adjustment or translation from the original form, the computer and robot make no adjustments. The form of a sculpture like *The Earth Laughs* is, in part, owed to a sculptor's capacity to use these tools.

The Broadway Blooms show coincided with the city's lockdown in response to COVID-19. I installed *The Earth Laughs* outside Lincoln Center with a crane, around midnight. A cab stopped at the light, and the driver rolled down his window. "Thank you," he said.

"Really?" I replied.

"Yes, thank you for bringing this beauty here to us."

I was totally moved and surprised by this expression of gratitude. However, soon I realized that it was not an anomaly.

After installing the show, I went down to New York City every couple of months to clean off the grime and soot that settled on the pieces—as well as any kind of vandalism that is typical. Usually when you put anything out in public, someone will do something to it. But I never found graffiti—no abuse, nothing. That was staggering. It turned out, the art had protection throughout the city. Whether in Harlem or Lincoln Square, no matter the neighborhood, people told me, "Don't worry, we'll keep an eye on this."

The whole experience has affected my artistic practice. I came to realize there's a relationship you're building with a form embedded with an emotional response. It's a strange and uncompromised gift.

That's what we hope to achieve in Chicago. You show up at 1000M and enter the lobby. Who knows why you are there? But you walk by these larger-than-life flowers by the window that I hope offer a moment of pause and welcoming and inspire a sort of rest and joy.

FRANCIS GREENBURGER: I believe that art can enhance the quality of a developer's building; I know it has enhanced my life. At fourteen, I bought my first painting—a scene of an elevated subway by the artist Ted Xaras that hangs in my personal office—and among the thousands of artworks I now own, it remains one of my favorites.

Some collectors are interested in art because of its historical value. They see a relationship between a work and an artistic period, and that link informs their viewing. Others with a more intellectual bent might favor conceptual art. My love of art has always been visceral. How art stimulates me, triggers my imagination, or sets my mood is the source of my lifelong passion.

In my twenties, I became friendly with the painter George Hofmann, who acted as something of an early advisor. A student during the height of abstract expressionism and part of the New York color field movement, he introduced me to his circle, which included Clement Greenburg—one of the greatest art critics of the twentieth century—and prominent artists of the time, such as Kenneth Noland and Helen Frankenthaler. George also introduced me to the work of less well-known people, who in the fullness of time turned into major figures in contemporary art.

My experience with and knowledge of art has grown over the years, but I still take an intuitive and very individual approach to collecting. I focus on less-known, sometimes younger but not always, emerging or midcareer artists as opposed to big names. My goal isn't to find the next famous artist. Rather, I enjoy supporting artists whose work I respond to and for whom I know my support is meaningful.

That support hasn't only been financial by way of purchases, but also exposure, in part thanks to Time Equities' Art-in-Buildings program, which brings art to the public areas of our buildings. When we first began in 2000, I was concerned the program might be an indulgence of mine. I discovered, however, that I wasn't the only person who liked art. In the last twenty-five years, we've received a tremendous amount of feedback that people know real art when they see it and that it reinforces the quality of what we are selling. A classic example is Travelers Towers, office buildings from the '80s outside of Detroit, that we gave some life to by filling them with art. "Some of the art is not to my taste," a broker who rented two large spaces said. "What my clients like, though, is that it shows the owner cares about the buildings."

As my art collection grew and was moving to different locations, I wanted to keep an inventory. I hired an art coordinator, who at some point put a letter in front of me to sign with the letterhead "Francis J. Greenburger Collection."

"What is this?" I asked.

"Well, you have a collection," she said.

That was the first time I thought of myself as not just someone who likes art but as a "collector." The art-coordinator position grew into a full-time in-house curator, which is unusual for a real estate company. Interior designers or architects are more than eager to pick the works themselves for a development project or bring in an art advisor. There are lots of art-advisory services, but the problem with many of them is that they are tied into the higher end of the market—and I'm proof you don't have to spend a million dollars for the art in the lobby.

Time Equities' in-house curator, Tessa Ferreyros, thinks holistically about the collection, considering artists I already have relationships with and throwing in some new names, too. Her recall is remarkable; I think she's memorized every work in my collection. When we are looking for something for a space, she immediately pulls it up on her iPad. It is one thing to have a piece of artwork, it is another thing to install it in another way that gives it presence—she is excellent at that.

TESSA FERREYROS

Director of Art and Buildings

Curator for the Francis J. Greenburger Collection

Sometimes you fall into your career. After graduating from the Rhode Island School of Design, I thought I would be a practicing artist but got a position at Sotheby's taking phone bids because I speak Spanish and French—thanks to having a Peruvian father and a French mother. However, I lost my job during the 2008 financial crisis and moved back to my hometown of Houston, where I joined the curatorial department of The Menil Collection. The thirty-acre walkable museum of nearly nineteen thousand works, founded by John and Dominique de Menil, who fled Nazi-occupied France for Houston, is where I really fell into the curatorial world.

Eventually I returned to New York for a position at the Museum of Modern Art and to earn a master's in art history from Hunter College. I later became a curatorial manager for the Madison Square Park Conservancy, a nonprofit serving the public space in downtown Manhattan. I knew Francis's former director for art and buildings, and when I learned she was leaving, I applied for the job of managing his private collection and overseeing the art programs across Time Equities' properties. I was attracted to the position because, although technically it's with a corporation, we are really running a mini public art program.

CasaMara is a great example. Because the West Palm Beach development was new construction, we had a chance to be involved in the process from the ground up and make the building more art centric. Not only did that mean making sure that the infrastructure and lighting showcased the art but giving Time Equities' Art-in-Buildings program a chance to exhibit part of the corporate collection. We learned that a private developer is required by law to pay a certain percentage of the construction costs to the city of West Palm Beach to install public art in the project. However, alternatively, the developer could install their own collection if it met certain criteria, including accessibility to the public at large. Through the company's Art-in-Buildings program, we installed *Caryatid No. 2*, a light-and-airy, yet physically commanding sculpture by Carolyn Salas at a busy intersection—and made CasaMara's core art collection available to anyone who makes an appointment to see it in the leasing office where it's exhibited. The entire initiative was very successful—visitors responded to it and the capacity of what art can do.

1000M was an opportunity for the same enterprise but on a much larger scale. I had my first meeting with Kara Mann about the scope and scale of the building in late 2019 but didn't start revving up until the floor plans began to be finalized in mid-2022. Until then, it was hard to have a sense of the look of the various areas and where the more important moments were for commissions.

Not to say that Francis hadn't been buying artwork during that period; he actively acquires works all the time. There is nothing unifying the thousands of works in his personal art collection—other than Francis. The form and content of the art vary greatly but all are a true reflection of him and his taste. Francis doesn't buy works for status or follow any kind of trends. He has his own eye, which is what guides him. Recognizing talent early on without needing it to be dictated to him, he often embraces emerging artists.

As 1000M continued to progress, I kept a database of works from his private collection that were possibilities for the tower so that when the building professionals were far enough along, we could begin figuring out what art would actually go where.

Construction of a building has so many moving parts. There is a huge team from the designers to the asset manager to the engineers. Everybody has important concerns. In that environment, how can you make sure that the art is also valued? A lot of it is advocating for the art's importance to the building and what the building needs to do

Eighth floor den with paintings by Kinga Czerska, left and Lilian Garcia-Roig, right

for the art to be a success. Because Francis understands how important art is to what he does as a developer, I always have his support.

Kara Mann, the interior designer for 1000M, and her team offered their thoughts on spaces for artwork, which began a lengthy back-and-forth until we finalized a plan of 140 pieces for the amenities areas and common spaces. This included art from other Time Equities' properties, such as three photographic prints on textured canvases we moved from 50 West to hang on the elevator landing of the eighth floor (the floor with the pools, game room, reading room, kitchen, and bar lounge). We also acquired art that purposely reflected a space, like the colorful abstractions for the sunrise coffee lounge. In spaces where the art had the potential for great impact—for example, the building's lobby—we commissioned pieces, which is much more involved.

No matter the process—whether it's pieces from his personal collection or commissioning a new work—Francis's level of investment in the property's art is meaningful. That's because he sees its art program as an amenity. Offering thoughtful contemporary art to tenants is another consideration of how the space will be most enjoyed.

FRANCIS GREENBURGER: There's a difference between buying existing art and commissioning new works, especially when you have a site in mind. A commission is very much a conversation and not unilaterally what the patron is asking for or the artist wants to make. Ron Klein's commission for 1000M's lobby is a good example of this type of dialogue. His work—large compositions of many small elements that he drills or nails to the wall—can take a variety of shapes and dimensions. Should he limit the composition to a more intimate scale or grow it up the twenty-seven-foot wall? This is just one of many questions that needed to be answered in a conversation.

This wasn't the first time I had worked with Ron to tailor one of his pieces for a specific site. I met the artist in 2006 after attending one of his exhibits, where I bought a sculpture. I intended the piece for my dining room, but it turned out to be too big for the wall. So I asked him if he'd consider shrinking the sculpture.

"Would you ask a painter to cut off part of his canvas?" he answered.

It took a few years, but I wore him down. I loved the sculpture and wanted to put it where I ate my meals. Finally, he figured out a way to make it fit, and through the process we got to know one another.

RON KLEIN

Artist

I was born in Chicago, so when the opportunity presented itself to build a sculpture there, I thought, "Gee, this is great." When I was twelve, my family moved to Highland Park. The suburbs, the land of surplus, influenced me and later my art. I discovered the world has enough stuff. In college at the University of Colorado Boulder, when I needed a TV, I went on the streets and tried to get to one before the garbage guys did. People throw stuff out that works. I got a television and discovered I was more interested in taking it apart to see what was inside than watching it. I realized I could reorder existing materials and see the beauty in objects already in place.

This realization was reinforced when I began to travel to extreme environmental habitats in the late 1980s. I loved to travel and wanted to see the world. Witnessing the untouched nature of the Amazon Rainforest or Madagascar's rainforests, I thought, "These objects are incredible. I can't make anything better than this."

I have spent the last twenty-five years collecting nature's throwaways from the jungle and rainforest and have a library of beautiful objects from its floor that I morph with industrial objects from farms in rural Pennsylvania. Essentially, I draw with objects. If I need a certain line, I use an object that is the line I'm looking for. Hundreds of small sculptures of different textures and qualities come together to

Ron Klein

create a larger visceral sculpture on the wall. I never want to be too specific because that locks you, and what I love about art is that you can look at the same thing and see it many ways.

When I'm making something, it's 100 percent about me; *here's what I want to say*. For 1000M, I wanted to create an abstract shape with some relevancy to the idea of a lobby. At the same time, I was thinking of children and all the different kinds of people that were going to be looking at the sculpture. I wanted it to be challenging. If you look hard enough, you'll see some commentary. A lobby is a place where you might pause. It's not a grammatical period. I mean, the period would be your condo where you rest. The lobby is an intermediate place you move through, what I think of as a comma.

I like riding that fine line between chaos and order. If a work is too chaotic, you can't look at it; but if it's too ordered, it's boring. This reflects my sense of the real world. We are all small parts of a larger

Detail of *Power of the Small*

whole and need relationships to grow—just as these objects are dependent on their positioning and relationship to one another and the whole to exist successfully.

As I was thinking about building the piece, I considered my own childhood in the Chicago suburbs and the condition of the world I currently live in. Growing up, I spent a great deal of my warm-weather time on a boat on Lake Michigan. Both calmness and turbulence reside deep in my memory and are reflected in the work. Its wind and weather are reflected in the movements within the sculpture. The ebb and flow of life is moving through the sculpture as it moves through our own day-to-day existence. Nature and industry are a fine balancing act that I hope come together in each of the objects.

Because this piece is in a lobby—where there is wind, fans, and all sorts of extra factors that wouldn't be in a collector's home or gallery—I had to construct all its elements for longevity. Not only did I need to make sure everything was really securely fixed, but I eliminated any frail organic materials that wouldn't stand up over the years. If this weren't a public piece of art, I might not be able to resist includ-

Power of the Small installed

ing natural elements that look raw and tough, even though I know they may not last that long. In this case, though, I went out of my way to select pods that are strong, clean, ready to exist in the world. I first create the sculpture in my art studio, drilling into the wall and attaching the objects with pins in a temporary fashion. This stage of my process is adventurous and experimental. Once I'm certain of the object's location, it's attached in a more permanent manner. It is really fun building it. My systems are nuts. Once completed, I take a picture of the finished project, and then when it's time to install the sculpture I project the image up on the wall, which becomes my road map. I get all the objects out of boxes, and recreate it. But the area has to be completely dark for the projection process to work.

The idea of blacking out 1000M's enormous lobby with light streaming in from the floor to ceiling wall of windows was the stuff of nightmares. I needed scaffolding, blackout curtains, a lift, clamp lamps, a projector, and a week to drill and attach 240 small and separate components in their exact position.

It's redundant recreating my sculptures, but I love that when they are on the wall, they look effortless. No indication of work—not of the adventures collecting the objects, the thought in their assemblage, or the effort in the installation—just the image.

I try to make each object special. It has to transcend its identity, or I have failed. If you immediately see an object for exactly what it is, then I didn't do my job. Preferably, when you get closer, then you might think, "I know what that is." The idea of mystery, throwing a curve ball at you (as the viewer), is important to me. If you recognize things too quickly, there's no challenge.

FRANCIS GREENBURGER: Almost a year after the initial meeting to discuss the artwork for 1000M's lobby, *The Earth Laughs* was installed on April 5, 2023, to join what would eventually be a group of works greeting residents and guests of the building. On the wall opposite Ron Klein's *Power of the Small*, we chose a work by Chin Chih Yang, a multidisciplinary artist who manipulates aluminum cans into woven abstracted tapestries in a commentary on conservation. Behind the reception desk—the focal point when you first walk in—are three works by Rakuko Naito, a Japanese artist I have known a long time. She makes beautiful and subtle monochromatic rolled-and-woven paper works that don't age. Their elegant staying power speaks to the ability to define space on a high-level without flashy or fleeting gimmicks.

The Earth Laughs' journey from New York to Chicago required two months of coordinating (not to mention the many months of planning by the architects and contractors to reinforce the floor beneath the 5,000 lb. of marble). On April 4, a team from Harry Gordon's art movers picked up the sculpture from Jon Isherwood's studio and drove it to Indiana. Because there were no locations in Chicago where they could park and stay close by the truck holding the sculpture, they slept at a motel two hours outside of the city. The movers woke up at 5 a.m. in order to arrive at 1000M by 7 a.m.

Loader picking up sculpture

JON ISHERWOOD: A knuckle-boom loader picked the sculpture up off the truck and deposited it on a roller. The guys at 1000M were incredibly helpful and enthusiastic about the program. They had to make sure the sculpture was in place before installing the revolving doors, otherwise it wouldn't have fit.

We drove the crane truck right up to the front of the building, where the sculpture was lifted off and placed onto large moving dollies. Then we had to get everybody to help roll and push it up the ramp into the lobby. At that point, the floor of the lobby was still a concrete layer, the marble flooring hadn't been done yet. Because of the weight of the sculpture, we didn't want to roll it on a finished marble surface.

We moved the piece into a position—there was extra foundation and structural work done to the base of the building where the sculpture sits—and then lifted it down from the moving dollies onto the floor. Once it was on the ground, I double checked that it felt right proportionally to the window, the doors, and overall space.

The Earth Laughs being sited

Back in New York, I'd shown the piece on a four-by-four-inch stainless steel base, but it was extremely apparent that that reflective material would be too brash and too harsh for the lobby. Instead, we chose a marble tile that relates to the sculpture's stone as well as the lobby's finishes. We left the sides off the base so that *The Earth Laughs* sat on a frame when we installed it. Then when the lobby was tiled, the tile went all the way up to the edge of the base. After installation, we shared a glass of wine and toasted the journey of the piece, from the marble that started out in Italy, to New York where it was crafted, to the artwork's final home in Chicago.

Chapter 15

THE DRAMA OF THE CRITICAL PATH, JANUARY 2023

Progress Is Made But Not Without a Few Delays

FRANCIS GREENBURGER: On January 25, 2023, I received a cheerful email from Randy Bullard with a picture of the rising tower. "To give you some perspective of what has occurred with the progress of 1000M," he wrote, "this is a photo of us pouring the matt slab one year ago this week and a photo of us now at Level 41."

Indeed, progress was being made. The concrete team was framing the forty-first floor with plans to be on the forty-third floor by the end of the month. The mechanical, electrical, and plumbing subs worked on levels ten through sixteen. Each trade has a different duration, so there is overlap, but the sequence always remains the same, with electric going first—roughing out the outlets, light switches, and more. RG Construction, the ceiling and wall subcontractor, was framing the walls and installing studs on levels sixteen and seventeen. The City of Chicago Department of Buildings was not far behind. Framing, plumbing, HVAC, electrical—everything has to be compliant, which is a big job for the city on a project like 1000M. Buildings came through regularly with a van full of people to inspect five floors at a time. Right now, they had approved the drywall installation for level five through ten.

That sunny list of accomplishments came to a screeching halt when problems with Americans with Disabilities Act (ADA) compliance came to light. The ADA issue arose nearly a month earlier when a sub asked for an RFI, a request for information, which is the process of clarification when a concern or question arises. In this case, the most difficult issue was that there wasn't a forty-eight-inch approach from the wall to the shower controls in about forty bathrooms on the floors already framed out. That's the floor space, which must remain clear for those who use wheelchairs to be able to maneuver. Issues always arise in the field, because drawings on paper are not the same as what happens in real life. This is not a perfect business, but there are no shortcuts or workarounds when it comes to building codes. Luckily, there are solutions to every problem. Solving them, however, always comes at a cost. There can be the hard costs of ripping out any existing work and redoing it. But even if it doesn't come to that, there are delays to the schedule and more change orders for the general contractor, all of which boils down to more money. Everything boils down to money.

By the time I received Randy's email, Jordan had been dealing intensely with finding the best (and most economical) solution to this problem.

JORDAN KARLIK

Principal, JK Equities

We don't want to install drywall on the floors that are already framed until we correct the issues, because it's easier correcting the framing than tearing off the drywall and then correcting the framing. We're literally talking about anywhere from a quarter of an inch to an eighth of an inch. But that's the standard these guys are held to. That's normal.

Beyond all the mumbo jumbo and technical drawings, the questions are: How do we comply? Do we switch the shower controls? Do we change the drywall? Do we just demo the whole thing and start again?

We are maneuvering within certain constraints—we have fixed pieces. Plumbing, like the shower drains and toilets, are cast in the concrete slab and can't move that easily. And it's got to be marketable and meet our design intent. We can't have a ridiculous-looking bathroom.

We're talking about moving the wall back, but is there enough room? Can we change the studs? What does that do to the fire and sound ratings? There's a whole cascading effect.

FRANCIS GREENBURGER: Jordan, the architects, McHugh, and the rest of the team went back and forth, making the necessary adjustments and tweaks to bring the building and *all* its bathrooms into full ADA compliance. But part of the "cascading effect" was that, by the time I arrived in Chicago in early March, the drywalling had been delayed a few months. McHugh was devising a recovery plan, which often means overtime. This was in contrast to the concrete and curtain wall, which were several months ahead of schedule. It is all part of the drama of the critical path. You need to stay on it day by day, because you never know what issues are lurking below the surface.

All the trades have questions that arise in the field. "Just seems to me that we usually have the most," the drywall subforeman, Jon "Woody" Wood joked about his trade. Drywall was just a shorthand for the much larger job of the wall-and-ceiling work by RG Construction, commercial wall and ceiling contractors that handle carpentry, framing, insulation, plastering, acoustical treatments, and more.

"Everybody puts their stuff in our walls," said Woody, who brought in over one thousand sheets of drywall every single week. "So, if I put a wall in the wrong spot, they put their stuff in the wrong spot, which is a lot of responsibility."

JON "WOODY" WOOD

Foreman, RG Construction

Growing up in Sauk Village, thirty miles south of downtown Chicago, I wasn't the best student and had a lot of energy. I ended sitting next to the teacher so she could keep her eye on me. Her husband was a carpenter, and she told him that maybe I would be a good helper. I started working weekends for my teacher's husband, John McGrath, digging post holes, carrying lumber, shoveling gravel—whatever I could do. I spent the summer after eighth grade working for him. He would pick me up, take me to work, and drop me off at home at the end of the day. I continued to work with him throughout high school. Mrs. McGrath and him took me under their wing. I learned the trade from him, and when I got out of high school, I decided I wanted to stay with it and be a carpenter.

For many years I worked with John and his brother, remodeling homes and framing our own houses. Residential carpentry is mostly wood, but you have to know a little bit about everybody's trade, like plumbing, electric, and fire safety. I always say I know enough to be dangerous. Then, in my late twenties when I was expecting my first kid, I decided it was time to join the union, to have great health insurance for my wife and kid—a pension, retirement, and such.

In 2006, I joined RG Construction Services, one of the largest wall and ceiling contractors in the region. When I began my first job on Trump Tower, I knew nothing about metal framing—and this was ninety-two stories of metal framing in downtown Chicago.

Carpenters are divided into residential and commercial. Trump Tower was a commercial carpentry project, even though it's a residential building, basically meaning that it's pretty much all metal. There's always a little bit of friendly competition between the two sides. We kind of don't like each other. One always feels like they work harder than the other.

Yoga class

Metal-framing buildings is a different kind of work than framing a single-family house in wood. When you're building with wood, you want everything as tight as you can make it. The tighter, the better. The more attention you can pay to that detail, making it perfect, the better the house. With metal, it's the opposite. A house is two, maybe three stories at most. Skyscrapers have to be able to move with the wind. You always leave tolerances, because the tighter you make it, the worse it is. It's a whole other type of construction that takes a little bit of getting used to (that and wearing gloves everyday so I didn't cut myself). But being a carpenter, I felt I could do anything. Anything they asked me to do, I did. I took my lumps, got the lay of the land, and figured out how to build things out of metal.

From Trump Tower, I went to 300 North LaSalle, a sixty-story building on the Chicago River, and became the guy laying out the building. As part of our scope of work, we measure and lay out where every wall is going in the building. We do the math, deducting space for drywall, finishes, whatever the case may be. Some walls get one layer of drywall; some walls get two; others need insulation.

Everything is laid out on the floor—every door, intricate ceiling, unusual details for amenity spaces. Generally, there are only a handful of guys on the job who have a blueprint. We lay that entire thing out on the floor so everyone behind us can see and follow without seeing the plans. Electricians, plumbers, everyone—they have to follow our framing, and we have to make sure their stuff fits where it needs to fit.

For 1000M, we had several crews working at the same time. This industry is quite similar to a factory line. Same guy comes in, does the same thing every day, and he gets pretty good at it. We have what we call our layout guys, who have the blueprint. They use the control lines to map out where the building goes. Then, you have the framers, who have a little more diversity in their job. They may put the tracks up on the ceiling off the marks on the floor from the layout team; stand up the actual studs that create the structures of the wall; or hang drywall. Our company also does the taping, finishes, and plasters.

There are a wide variety of materials in 1000M: high-impact board that handles damage a little better and mold-resistant board in bathrooms and any wet area. Providing walls that carry certain fire ratings—like a two-hour fire rating, designed to resist fire for at least two hours—that is part of our scope. This building has a lot of sound-rating properties as well. In an apartment building, you don't want sound to travel from one unit to the next. There are several ways to achieve a good Sound Transmission Class (STC) rating. A chase wall is a gap or separation between two walls, usually used to run plumbing, electric, and heating. A number of layers of drywall helps change that STC rating. In this type of building, every single wall is caulked and rated for how much sound it leaks. Shiner Acoustics is a company that everyone in Chicago uses. They'll come out and go

Kids playroom

into a unit, make noise on the other side of the wall, and measure it with certain devices.

There are a lot of different finishes in 1000M. Sometimes it's a combination of drywall and skim coat or cover coat. We provided a finish of a thin layer of plaster on the exposed concrete ceilings. Most people probably wouldn't even notice, but it is a specialty finish. We pride ourselves in every job. It's a craft. And a race. But I've been doing it for seventeen years now, so to me it's like walking.

In a project of this magnitude, there are always challenges and surprises. We all had to collaborate with each other to make sure we met the requirements. This building is state of the art with beautiful finishes, amenities, and artwork that is elegant and simply amazing. Working with designers and architects who have a vision and passion to create a quality building that will live in the Chicago skyline for many years to come was a great experience. I am proud to have my name associated with 1000M.

FRANCIS GREENBURGER: As a developer, I make it a practice to periodically visit any properties we are working on because, no matter the technological advancements, there is no substitute for experiencing the physical space and meeting people face-to-face. My March visit to Chicago was scheduled around the final design meeting for the all-important amenities.

I had asked for a wholesale revision of 1000M's eighty thousand square feet of indoor and outdoor amenity space spread throughout the tower. There is all this thinking, rethinking, double-checking, and then making changes where appropriate. The process doesn't stop; you keep going until the end. If you see a mistake, better to change it, or else you have to live with it until the end of time.

The original proposal included a second, separate floor of amenities on the twentieth floor for the tenants of the building's more expensive units. Rob opposed creating two classes in the building, and I agreed. Often, the market doesn't respond to these "exclusive" zones as strongly as experts think they will. The mere fact of their existence inherently designates the other part of the building as "second class"—and in this case that would have been two-thirds of the building. I wanted the *whole* building to be a luxury environment. We asked all the teams—the designers, architects, developers, rental consultants, and leasing team—to go back to the drawing board.

The result was a skyscraper with more amenities space than any other residential building in Chicago. 1000M has a seventy-third-floor rooftop deck (the highest in the city and accessible to anyone in any apartment); a two-story fitness center, including a basketball court; an entire floor dedicated to coworking; private event spaces and social lounges; rooms for toddlers and teens; a small art gallery; and guest suites for out-of-town visitors. It's a whole world within a building.

The eighth-floor amenity level is home to the indoor lap pool and spa and outdoor pool and deck. Installing the pools was a milestone. A unique feature with its own permit, the outdoor pool had its own evolu-

Basketball court

tion that began in the early days when I said it should be usable during winter. The room went insane. In Chicago, outdoor pools sit covered for nine months of the year. They thought I was a maniac, but in Colorado, where we take family ski vacations, every $50 motel has somehow figured out how to keep a pool open in the snowiest of winters. Fast forward through *lots* of meetings and negotiations, we extended the season thanks, in part, to glycol (a viscous fluid that doesn't freeze) tubing running under and heating the deck. Although, Phil still insists, "In the dead of a Chicago winter, people will have to swim inside."

GREG CARNFORTH

President, Chester Pool Systems

Chester Pool Systems has been in business since the 1950s. Located in Southern Indiana, the company and its eighty-five employees fabricate pools and ship them, either complete (up to sixteen feet wide) or in sections to be assembled on-site with our own forces.

We were contacted by the McHugh team very early in the building process. We have installed dozens of pools with McHugh and find them top-notch. All of the lofted pools are a challenge, but McHugh is very detail oriented and made sure Chester had a voice at the table early on. We assigned senior project manager, Andy Thompson, to the project, who pulled together the engineering, purchasing, and shop and field departments to ensure that the build followed procedure. This was a fast-track project, with the product needing to hit a schedule that was planned months in advance. The shipment was over width and had to be permitted by the day into the city.

We only build lofted and rooftop pools from stainless steel, which is 75 percent lighter than traditional, poured-in-place concrete pools and requires less structural support. We built out the $4 million aquatics package for the Revel Casino Hotel (now called Ocean Casino Resort) in Atlantic City, the Margaritaville Resort pool looking out over Times Square, the Four Seasons Hotel Boston, and over fifty high-rise vessels in Chicago.

It is challenging to put a pool in a building. But go to any hotel, condo, or luxury rental website, and the pool is on the home page.

FRANCIS GREENBURGER: When Phil, Jordan, and I gathered in Kara Mann's brightly lit office in March for the design meeting, it was the culmination of a lot of mood boards, visual presentations, and trays of materials for us to touch and contemplate. Despite all the previous debates, there was still a lot of territory to cover. It takes a lot to decorate eighty thousand square feet.

Floor twenty-one had become entirely devoted to work, which meant private offices, an open coworking lounge, a "focus lounge" (aka a meditation room), Zoom pods (complete with acoustic walls, so you can't hear your neighbors, and ring lights, to look your best on cam-

Top: Pool under construction

Bottom: Finished pool

Drink lounge rendering, palette and plan

era), and conference rooms. We settled in to discuss everything from the overarching palette to the signage for the numbers outside the individual office spaces.

In the winter, the designers had offered a look that I thought was too dramatic for any kind of office.

"You have a lot of drapes going on in the other areas," I said during our December meeting. "I know that is one of your signatures, but I think this should say 'workspace'"

To that end, I asked to see an alternative to the green vinyl chairs they picked out.

"Color wise it seems a little funky. The other thing that is concerning—it is very slippery so you could be sliding around on the chair," I said. "Every chair has to be sat in by one of you and the comfort has to be attested to. Chair comfort is important."

I might have come off as difficult, but I felt it was important to reinforce practicality, even if it forced Kara Mann's team to alter their creative vision. Earlier on, they presented a scheme that included elements such as burgundy-painted millwork; wallpaper with faces, coffee cups, paper clips, and objects; and a burgundy-and-cream patterned rug for the coworking space. There was, as I commented, "a lot going on."

Eighth floor cocktail lounge featuring *Girl in the Pearl Necklaces*, 2021, by Frances Goodman

"This is a workspace, not an evening space," I said. "A heavy visual experience distracts from the work you are there to do."

After a healthy debate about the purpose of color and pattern, they agreed to move to a lighter palate. I guess the customer is always right. When Kara Mann's team presented the final blond-wood flooring and doors and white-dove walls in March, they still found a way to bring in color through details such as deep-blue furniture and dark millwork and countertops in the coffee and print areas. There was some drama to the darkness, and while I wouldn't want it everywhere, I thought they had finally struck the right balance. Of course you can't make everyone happy. "Everything shows on a dark countertop," Jordan said. Kara Mann's studio lead, Laura Lee McAllister, took a note and suggested they find "something with a little fleck."

On to the next floor, level 20. Home to a game room, playroom for younger children, a bar, and private party lounge and terrace, this was the opposite of the workspace floor above; and yet I felt the first design plan was "too plain."

Harbor Bar & Lounge

“What is the sense of elegance we are projecting for this luxury floor?” I asked.

They returned with a palette that I thought was comfortable *and* elegant. The Harbor Bar and Lounge overlooking the lake and park was a knockout. The entire floor was a cool combo of boxes of honed black tile surrounded by frames of polished white. Behind the bar, made of Kara Mann’s trademark black marble with dramatic white veining, is the visual inverse: a smoked mirror in a white-lacquer frame; and a gray-blue special-plaster finish on the wall ties up the room in an unexpected but not jarring element. I questioned the sisal area rugs underneath the low lounge seating by the wall of windows looking out over the lake. “Sisal is a significant contrast to the other elements,” I said. “The simple and obvious thing would be to go a darker burgundy. You are bringing in a certain level of informality.”

“Yes, that’s our intention,” Laura Lee said.

Although I didn’t 100 percent understand it, I was okay with that decision. (Area rugs are easy to swap out.) But I did have feedback on the sofa in front of the lounge's fireplace: It was too short compared to the couch on the other side of the room. You want balance.

"Good note," Laura Lee said.

The meeting continued from room to room. I objected to the skirted chairs chosen to "tone down the masculinity" around the poker table in the Sunset Lounge. "At a poker table, you need to be *at* the table," I said, asking for chairs that were easier to maneuver. They figured out a way to solve a long-standing problem to get sixteen seats around the main dining table in the Private Party Lounge. I expressed my true feelings about the lighting fixture above the table but didn't push it: "I don't know who came up with the concept that a pipe with a lightbulb on either end is a chandelier."

We ground our way through so many details: sconces for a library feel; the tone of the wood for furniture feet; the commercial-grade corduroy of a sofa; the backsplash in the coffee bar. Among the differences of opinion was a burgundy. "Art does not want to be on a purple wall," I said, in my attempt to bring as much fine art to 1000M as I could. "There's a reason that all art galleries are white."

"We are trying to make this room moody," Laura Lee pushed back. "I can't pretend it's an art wall."

"I would do the wall white and maybe we can find a burgundy painting."

"That's the final floor," Laura Lee said, sighing deeply.

"Does this mean that we have to build another building?" I joked. We both knew full well that there was still a lot more work before finalizing the plan and welcoming tenants into the finished product. Whether it's ADA compliance or a fixtures company that has gone out of business, issues always arise in the field.

"Good luck getting free of this project," I said.

"It will be nice to have a weekend," she replied.

My attention to, and interest in, design is not the norm for a developer. I have a strong aesthetic and logic, which is a big part of why I

Co-working lounge

decided to do a project like 1000M. I believe that the value proposition is to create something of extraordinary quality. If I produce something that is 5 percent cheaper but doesn't have an expression of worth, the market could pay 30 percent less. CasaMara is the talk of the town in West Palm Beach and making incredible returns I believe because of the thoughtful design and sweating the details.

Leaving Kara Mann's office, I wondered whether my input made any difference in the end. If they were left to their own devices, would they come up with equally good solutions?

Maybe I'm indulging myself because I like to be part of the design process. In truth, I think we all benefit from someone pushing back on our assumptions. It's like a writer without an editor, or actor without a director. When it came to the design of 1000M, hopefully I made a contribution.

Shuffleboard table in the game room

Chapter 16

PINS AND NEEDLES TIME, AUGUST 2023

As 1000M Tops Out, Budgets Tighten

FRANCIS GREENBURGER: Topping out is a big moment in the life of any building. A rite marking the symbolic completion of a structure, it can be putting the last beam in place. For 1000M, the end of the vertical phase of construction came when McHugh poured the final concrete floor on August 4, 2023. Since the mat slab in January 2022, a total of 54,791 cubic yards of concrete had been poured to create the bones of the seventy-three-story tower.

The rest of the trades followed from the ground, moving floor by floor; legions of tradespeople framing, drywalling, skim coating, installing appliances, taping, and painting. Among that army, were the elevator subs, who had begun in late February to install the elevators that would eventually zip up and down all seventy-three floors.

Elevators are a tricky business. At 50 West, we had a long, terrible period where the elevators kept stalling. In tall buildings, elevators are very sensitive. There are many security features to make sure that nothing goes off the rails—literally and figuratively. Sometimes, however, technological advances and reality clash. The systems at 50 West turned out to be *overly* sensitive. The very involved saga wasn't resolved until we flew the inventor of the system himself from Japan to fix it. It was quite an education in the intricacies of tall-tower elevators.

Perhaps it's ironic that Otis—the company providing 1000M's elevators—began when founder Elisha Otis invented the safety break. Hoist systems have been around since ancient Rome, but none of them could lift and lower people or materials securely. A tireless tinkerer, Otis devised a system of toothed, wooden guide rails that fit into the opposite sides of the elevator shaft, so if the hoisting cables that lifted and lowered the cab broke, the release of tension would throw a spring mechanism on top of the elevator into the notches and stop the free fall of the elevator and anything inside. Otis enlisted none other than P.T. Barnum to debut his invention at the 1853 World's Fair. In front of a crowd, Otis stood on a platform hoisted into the air. Spectators gasped in horror when the supporting cables were cut. Instead of crashing to his death, the safety brake activated, stopped the platform, and a new way to travel through the world—vertically—was born.

TODD BERGEMANN

New Equipment Sales Representative,
Otis Elevator Company

In the elevator industry, there are three different areas of work: one of them is servicing existing elevators; another is modernization; the third—my focus—is new equipment, whether it's an existing building looking for a brand-new elevator or a brand-new building looking for a brand-new elevator.

Before a company like Otis comes on board, an elevator consultant, typically hired by the architect, assists with designing the appropriate system for the building. The first task of the consultant is to generate an elevator traffic study, which takes into consideration the height, use, population, and type of the building. Is this a luxury condo or mixed-income housing? Office space or a hospital? Understanding the building is crucial to providing the right quantity of elevators and determining the speeds at which they will run.

After the consultant has given the specs for a building's system, the general contractor reaches out to different elevator companies, like Otis Elevator Company. In assembling pricing, we consider not only the architectural building plans but also the specifications for any kind of vertical transportation such as escalators and elevators. If awarded the contract, like any other sub, we then start the whole process of generating shop drawings for approval before getting the elevator equipment into manufacturing and, finally, delivering them on-site for installation.

In a perfect world, the elevator cab would feel like it's floating. However, that's not the world we live in. The elevator system is designed to operate in plum conditions, but a tall tower is not a perfectly rigid structure. It's designed to sway back and forth. An important topic is what happens to the elevator system when that building starts to sway and making sure its design mitigates the tower's motion.

Above the cab are steel-core cables that go into the machine room and come down onto a counterweight. Cables also hang underneath the elevator cab that go all the way into the elevator pit and back up to the underside of the counterweight, essentially creating a balanced circle. When the building moves side to side, so do the elevator cables. To counteract the sway, we add mass. If you add more weight, then you need a bigger machine to move it. The other consideration is if the building sways to extremes. Safeguards are necessary to shut the elevators down, so they don't do damage to the building, or, most importantly, the people inside.

At Otis, we pride ourselves on ride quality; so, another issue is vibration. The higher the elevation and speed, the more passengers can feel any imperfections in the system. One of the bigger factors in ride quality is the alignment of the guide rails running from the top to the bottom of the building. The elevators should *not* be riding on the

Harry Kolk, Otis Elevator

guide rails. Instead, their purpose is to stabilize the cab in its travels. But for that, they must be perfectly aligned. The other major factor is getting the correct balance between the cab's mass and its counterweight on the other end of the cables. If the whole system is in balance and the guide rails are perfectly aligned, the cab does really float smoothly throughout the hoistway.

We install some components in the elevator shaft while the general contractor is still building the top of the building. Those components include the guide rails, counterweight rails, and entrance doors.

1000M's tower is serviced by our SkyRise elevators, which run from a big heavy gearless machine in a dedicated room at the top of the building where the brains of the elevator system reside. In the elevator world we talk about capacity in terms of pounds. There is a service elevator rated for 4,000 lb., and the rest have a capacity of 3,500 lb. each. In a perfect world, when these elevators are traveling at their speed of 1,200 feet per minute, you don't feel them moving at all.

FRANCIS GREENBURGER: At the time of 1000M's topping out, the curtain level was up to the sixty-seventh floor, and Brandon Tobolic was part of the crew of ironworkers installing the glass facade. More specifically, Brandon and another ironworker stomped on top of the metal frame of the window panel with both feet to ensure they were locked into place. In window installation, an impeccable seal is vital to keeping out air and water. Even though they wore safety harnesses, it was terrifying to watch them jump up and down on a narrow piece of metal at a dizzying height. Brandon, twenty-one, had only a month left in his three-year apprenticeship to become a journeyman in Ironworkers 63, the union of Architectural Ornamental Ironworkers. Keeping a close eye on him was the boss, who also happens to be his dad. Steve Tobolic, in charge of 1000M's window installation, encouraged his son to enter the trade after high school. "He thought it was a good idea, so I

CK2 window install

Topping off

started working in a company shop, and six months after that I became an ironworker."

"I followed in my dad's footsteps," Steve said. "I thought Brandon would probably like to do the same as me. So far so good."

As to what qualities make a good window installer, Brandon said, "Trying your best and showing up on time to work every day."

"And learning from others," his father chimed in. "I think that's the key. Learning from your peers is how you really improve your skills."

Brandon found juggling his classes while working on the building the most challenging aspect of the job, much more than the installation itself. "This is curtain wall, so it's simple steps," he said. "It's like LEGO."

Bouncing around on the edge on one of the tower's top floors, it was clear he didn't have to overcome a fear of heights. "I've never feared heights," he said. But did his dad have a fear of his son *not* having a fear of heights?

Topping off

“I do every day,” Steve said, “for himself and anybody that’s on the job. It’s always in the back of my mind. Our company and all other ones on this site do a great job with safety protocols. But the fear is always there. And pride. When Brandon gets through this, he can look back at what he’s done and possibly show his kids one day.”

That day, for Brandon, was as distant as his perch atop the skyscraper. He still lived at home, commuting back and forth together every day. They never talked shop, however. They usually didn’t talk at all, because, as Brandon said about his dad, “He’s always looking at his phone.”

On a project like 1000M, the details are endless. A human’s touch could be found behind every window frame, concrete slab, light switch, and throw pillow that would eventually make up 1000M. The pillows, which Kara Mann’s team specified for the amenities areas, were of particular concern to Shirley Loenneker. “So many pillows!” said Loenneker, the project manager for the procurement company hired to purchase every item and make sure it arrives when it’s supposed to. “Everything has its own purchase order, tracking, and lead time.”

SHIRLEY LOENNEKER

Senior Project Manager, The Ness Group

I went to school for interior design and, after graduation, connected with a woman who was starting a procurement business and needed some assistance. I got hired to do project management on contract and never left that side of the industry. Or as I like to say, I fell ass-backward into project management.

I really dread when people ask me what we do at a procurement company, only because it's difficult to summarize—at least at our company. The lion's share of what we do is establishing and building relationships between clients and vendors to bring about a harmonious product among all the parties. It's a lot of schedule management, budget management, expectation management, and communication. So, it's definitely a lot more than, "Here's a list of twenty things. Go buy it and call it a day." Although, purchasing is a big part of the job. The longer I stay in this business, the harder it is to find something I *haven't* bought. I have purchased everything from swimsuit spinners to arcade games.

By the time we're brought on by a client rep, ownership group, or developer, a good amount of the design development has already been completed and established. We're handed a book of specifications of everything design has selected, and ownership has reviewed and approved. Then we send all of that out for pricing and bidding. Typically, we only bid out items that are custom pieces. If a specific item is ordered, you're going to get that item unless we need an alternate for time or schedule. For example: if the client wants seventy-two designer chairs, we go to the designer. However, if the designer comes back with $5,000 per chair as the best trade pricing and a thirty-two-week delivery time frame, neither of which would work for most clients, we make suggestions on less expensive alternatives in a tighter time frame. For the custom pieces, we reach out to whoever

Topping off

we have a relationship with at a company, whether that's a project manager or sales rep. Again, it's about relationships.

Once those choices are all settled, we move along to purchase orders and expediting. Then a lot of our job is just making sure everything gets where it's supposed to be when it's supposed to be there, which is a Herculean feat, especially when you're dealing with a lot of custom items. For 1000M, I think our fabric count came out to be like 150 separate line items. And for every fabric that you get, there's a cutting that needs to be sent, reviewed, and approved, and then the approval needs to be sent to the vendor. For every piece of custom furniture—table, chair, sofa, and so forth—you need finished samples of every different finish that's on the piece. If you have a table that has a wood base, a stone top, and brass-edge detailing, that's three samples. It's a lot.

The thing that can really get away from you with the custom items are the lead times. Once everything is in and approved, it's a lot of herding cats. When we release our purchase orders, we ask that the vendors give us a critical-path schedule that commits them to a specific time frame for everything from shop drawings to transit time. Revisions can become a problem where, at some point, a compromise needs to be made: either design needs to approve it faster or the vendor needs to make it better and sooner. That's where me and my charming nature come in to inquire: "Hey guys, what's happening? Are we ready to go?"

The challenge with 1000M was just the sheer amount of individual line items. The size was definitely not an anomaly for us. But you can't compare a project like 1000M to a large-scale hotel. If you're working on a 250 or 500 key hotel, that's a decent number of rooms, but you probably only have about five room types with the same basic stuff in those rooms for which you're just ordering a high quantity. The hotel might look larger than 1000M, but this is big in its own right. Like I said, there were so many pillows alone! We had up to two hundred purchase orders with some of them having forty items on the order. It's just a lot of individual little beasts to keep track of.

We have procurement software we use to enter in the specs, move to requests for quote (RFQs), then purchase orders, expediting, and so on. Every week we send out reports to the client of where we're at with the orders. I also break down a high-level bullet point of anything that needs massive attention. I remember Jordan [Karlik] telling me, "I can't with this report. It's like seventy-something pages!"

Unfortunately, I can't push a button and create a document that has the percentage of the POs or deposit issues. We don't have a workflow like that. Instead, it's the project coordinator, me, and the way our crazy brains are wired to keep track of everything. Whether we admit it or not, I would say a good amount of us have rampant ADHD because you just need the ability to both hyperfocus and multitask at the same time. Every week, the project coordinator goes through and touches every order. We run through the system by purchase order, look at the last time we've spoken to the vendor, the date they are supposed to deliver the item, and hope to God you don't miss anything. I run on a lot of panic.

With every new project, I ask myself, "OK, what have we learned? How can we do this differently? How can we do this better?" And every project is different because what works for a 1000M is probably not necessary for a small restaurant in the Bay Area. I'm always trying to figure out the new thing that's going to make this one run more smoothly than the next. But those things are often quickly abandoned when the panic sets in and we revert to the tried and true: "Just get it done!"

We have a high level of autonomy at our company. We're all trusted to get our jobs done, because if someone isn't, it becomes evident really quickly. We work very closely with each other. You're going to be on the phone with your project teammate constantly, poring over the details of these projects. You end up close to them no matter what.

While I start with anywhere between 70 and 150 emails at the top of my day, I can't even quantify the number of phone calls I make in a day—except to say: a lot. Honestly, it's kind of bonkers. My headphones are always dead at the end of the day.

FRANCIS GREENBURGER: When a building tops out, budgets start to tighten. At this point, you have typically begun to use your contingency, so if anything goes wrong, there is less flexibility. The fear is: Can you meet the project without exhausting your contingency. Otherwise, you are over budget. It is all pins-and-needles time.

Rob completed a comprehensive review of the budget, taking a look at the hundreds of change orders on a project like this. We seemed to be okay on the two most important elements in the industry: budget and schedule. There were some discretionary change orders, like Rob's closet upgrades, which was a couple million dollars upcharge. Then there

Roof garden installation

was the whole ADA issue, which was another couple of millions of dollars to fix. It adds up quickly.

The biggest unforeseen expense, however, was the increase in interest rates. Well, it wasn't totally unforeseen. Interest rates are always an unknown variable in any project, but with 1000M they were taking on a large position. In late July 2023, the Federal Reserve raised its funds rate to a target range of 5.25 percent to 5.5 percent, ostensibly taking borrowing costs to their highest level since 2001. The hikes that had been happening all year were the Fed's attempt to slow inflation, which threatened to reach levels we hadn't seen since the 1980s.

Fortunately, at the start of 1000M, we bought an interest rate cap. This is an insurance policy that pays you if and when interest rates go up. In other words, during the first two years of construction, they can't go higher than a certain number before the insurance kicks in. By February 2023, the insurance had kicked in. It cost us around $1million for a policy worth roughly $10 million, that would take us through mid-2025. For fifteen months our interest rates were very low, basically half the current rate due every month, with the rest paid by the rate cap. As we approached fall 2023, that was looking really good because, according to the press reports, the Federal Reserve wasn't cutting rates anytime soon.

If we didn't have the rate cap, we would have been over budget in interest, which thankfully we were not. That didn't mean our worries were over. The question then became what would the rates be in 2025? As developers, we have to think not only about the interest costs of the construction loan but also how to pay it off, which in the case of 1000M was $305 million that was due to Goldman Sachs on December 1, 2025. At some point, we want to get a permanent loan. The timing of that is very critical—you don't want to be stuck with some high rate.

Looking at economic models that attempt to predict interest rates, they basically showed interest rates starting to fall in 2025. Therefore, my expectation was that we would be refinancing the project sometime in late 2025 or perhaps early in 2026. The refinancing into a "permanent mortgage" would involve fixing the rate for the next ten years. Therefore, waiting for an advantageous rate environment was critical, but the building also had to be complete and substantially rented. However, while completing construction and waiting for lease-up to be completed, we were concerned that we might face higher than budgeted interest costs. To balance the risks, I decided to spend over $2 million on an insurance policy called a rate cap that would pay any interest costs over the budgeted level. When rates climbed even higher than anyone imagined, the rate cap became a lifesaver.

I wasn't the only one concerned about interest rates. Back in March, I had lunch with one of the guys from Goldman Sachs. When interest rates are rising, the reserves that banks have built into construction loans become inadequate, and they ask developers to put up more money. If the rate prediction is no longer true over the course of a loan, the loan is "out of balance" and only balanced by more equity.

When it comes to construction loans, typically more things go wrong than right. In this way, 1000M was looking good as we moved into the leasing phase of the project. We had probably used up around 70 percent

Roof garden installation crew

of our contingency, but that is not abnormal for a development, particularly one of this size. But it didn't take more than looking out from level 47—what Phil calls "the prairie view," because it goes on forever. All around us were cautionary tales: developments "on ice" or dreams for projects that will never get done.

Costs and income are changing constantly depending on the market, so we had to do sensitivity analysis regularly. Rob and I did budget checkups on an ongoing basis, as it's very important to keep checking your budget alignment. He also did a pro forma on rents. When we started the project, we predicted $4.60 per foot. At the beginning of the summer, the prediction was $5.10 per foot—15 percent higher. Interest rates were higher than anticipated but so were rents. There is usually good news in one place and bad in another, but if you budget conservatively, you come out with the right bottom line—even if the scenario is different than the original.

As to the upcoming leasing push, we were feeling good about the landscape, our product, and the team. The whole strategy is how many leases can you sign up as quickly as you can. We renovated the leasing center on the ground floor of the building we owned next door to 1000M, tearing out all the condo finishes from the vignette area and reprogramming it for rentals (another cost). As Rob said, "A normal person would have gotten rid of the whole sales center and rented it out for income." Instead, I decided that when the building started to open, it would be good to have a leasing center on the street level.

While 1000M's construction was moving along, the market seemed to be strong for renters. *The Real Deal* published an article on August 30, 2023, titled "Chicago leads nation in rent growth." The real estate trade publication reported, "Rental rates in Chicago have risen 3.6 percent since August 2022, tripling the national average of 1.2 percent and exceeding twenty major metropolitan areas."

We were also feeling good about the interest in 1000M specifically. The people who had registered online for an early look were somewhere between eight hundred and one thousand signups. No prices were listed on 1000M's landing page, so some of those people might have been delusional. But it did indicate interest as we entered full marketing mode.

The new sales center looked good and was filled with leasing agents, who had begun rehearsing their sales pitches, alternating between the role of customer and agent to be able to give one another feedback. In real estate circles, 1000M was the talk of the town, the object of a lot of positive buzz.

Expectations were high. The bankers seemed very pleased. But a big unknown drama was still set to unfold, day by day, over the next six months. Consumers might have loved the building, but how would they react to the pricing? That was the big question right now: Would we achieve the rents?

The team was very confident, but reality often tells a different story. May 1 was the date when the first tenants moved in. We would see.

Chapter 17

"DON'T TELL US WHAT YOU THINK WE WANT TO HEAR—TELL US WHAT YOU THINK WE SHOULD KNOW," JANUARY 2024

Leasing Up

FRANCIS GREENBURGER: These days you build a building twice. First digitally, and then in reality. And if you think the digital imagery is simply the work of a few mouse clicks, think again. It's almost harder to create the computer renderings than construct the actual building.

Rob was knee-deep in the digital marketing of 1000M in summer 2023, with the July 10 launch of a teaser page, 1000MChicago.com, which invited potential renters to join a VIP list for priority leasing-office appointments, unit-layout previews, and early access to pick out an apartment. The major draw of the website was the digital renderings of the interior and exterior of the tower, including a stunning aerial view.

Seventy-third floor rendering

ROB SINGER: We spent two years working on the project renderings—3D images of the exterior building, as well as unit and amenity interiors. We worked with Pictury, a rendering company based in Spain. One of their employees, Albert Neira, reached out to me by email after seeing the low-quality 3D image of the exterior of 1000M published in the media when we announced the switch from a condo tower to rentals. He wasn't impressed. I don't typically respond to cold calls, but in the world of rendering studio houses, Pictury is known to do really good work. Albert had thoughts. "Get serious about your renderings," he advised. He was right.

That started a long relationship with Pictury. For every rendering—of which we've done upward of seventy-five—there were many steps to the process that proved challenging to coordinate.

The images were worth the time and investment. Beyond the marketing aspect, they helped us really see the building and the spaces prior to construction, allowing us to understand our designs better and make changes if necessary.

ALBERT NEIRA

Marketing and Project Manager, Pictury

Renderers are the real artists of the twenty-first century. There are probably over one hundred studios out there creating renderings, but not everyone who works with the real estate world has artistic sensitivity. There are a number of variables that define the high-quality digital studios. First and foremost, there is a very skilled workforce with different backgrounds in architecture, fine art, and interior design. Not many companies are willing to put in the time to take an image from zero to one hundred. At our firm, one image takes, on average, about two weeks to produce.

In any project, the client first lets us know how many views they want—and what type of views. Whether an image is eye level, semi-arial, or arial, each artist is carefully selected for each view type. Arial requires the most, an order of magnitude more than eye level, because there are so many buildings and details to take into consideration.

For each image, we start the process by considering the lighting and angle, because those have the greatest effect on the image's mood. The angle is wherever you choose to put the camera, while the light is determined by the time of day (i.e. sunrise, sunset, night, dusk). With those basics determined, there are four main requisites to any architectural rendering:

1. A final 3-D model of the building or unit pictured from the architect: This may sound obvious, but if an architect updates the model, it will kill the project manager, and then the assigned artist to the concerned view, who will have to redo *everything*.

Digital rendering of an apartment

2. Landscape and material mock-ups: This is what goes in and around a room or building—what furniture or flowers you want in place—and requires discussions with the client, interior designer, architect, and landscape architect. The client will provide reference images, if the item is from a specific brand. But we have huge libraries of different chairs, tables, and other reference images. Sometimes we model furniture. However, if it is extremely specific, typically the interior designer creates a 3-D model. Regardless of where the reference image comes from, we place it in space, turning a chair fifteen degrees to the right or honing in on the tops of boxwoods.

3. Reference angles: Clients set the camera's angle toward what they want the viewer to see. It's a very simple task, yet it determines the ingredients an artist will use once they start working on the assigned view.

4. Google Maps location: This is so important we don't start without it. The location determines the human and natural elements. Rendering projects all over the world, we make sure to use culturally appropriate features. A Dubai project is different than a Chicago project.

At Pictury, one 3-D artist works on one image at a time. They convert and optimize the 3-D models that come in from a variety of sources in various types of software are converted into Autodesk 3ds Max software to create two-to-three different mood variations for the client to choose from. Once the client picks which variation they prefer, the artist refines the image and sends it back to the client for notes, which they will eventually use to complete the final image. The process for an image typically takes two weeks in total.

1000M, however, was not a typical project. I understood Time Equities wanted to make history. Usually, we hire a drone partner to get the background imagery for the building. In the United States, the FAA (Federal Aviation Administration), which regulates drones, doesn't allow them to fly higher than four hundred feet. When you place a camera angle above that height, you need a helicopter.

I joined Pictury in 2019, but 1000M was my first time managing a helicopter shoot, which is extremely expensive and precise. To insert a building in a seamless photo integration, we have to get the specs of the camera angles down to the millimeter. Although we sent our best photographer from New York to Chicago, I was so stressed that I was literally sweating while the helicopter was up.

I wanted to get the very best pictures and meet Rob's expectations. So, when he came back to us that year for another set of arial views and twenty-six interior views, I felt I had really accomplished something of beauty and importance.

Working hand in hand with Rob and Phil Castillo, there were a lot of comments. We couldn't be strict with our timeline. The amenities

Exterior rendering

areas were particularly challenging because marble tables are not an area of our expertise—especially not custom ones. It took about a month for Kara Mann's design team to produce models of specific furniture for us. Bumps and all, Rob and Phil wound up happy with the quality—so much so that we were hired for another forty-one views for a total of nearly ninety views. It was a huge project.

My favorite images are from the first helicopter shoot. Taken at sunset, it was so well accomplished because you have the blue of the lake, the purple pink of the clouds, and the gold of the sun reflected in the tower. When I saw those images, it was glory itself.

FRANCIS GREENBURGER: In the first week the teaser website went up, we got 250 registrations. That might sound good, but you're lucky if 5 percent of the people who sign up for a premarketing campaign actually sign a lease. We had seven hundred units to lease. (The other thirty units are affordable units or guest suites.) That was a lot, and deadlines for getting them done were approaching quickly. We were on schedule to have the units on floors one to seventeen done—and hopefully occupied—by May 1, rolling out more units each month until November. With the leasing center opening in December 2023 and the construction loan needing to be repaid in December 2025, we had given ourselves two years to lease-up the building and—again, hopefully—attract a favorable permanent loan.

The problem was that the traditional leasing season in Chicago is between April and September due to the warmer weather and the fact that people don't like to move during the city's colder months. The vast majority of leases expire in this time period, so the trend perpetuates itself. This is when most people make their decision about whether to renew their leases or look for a better alternative. Some leasing does occur in the fall and winter but far less than in spring to early fall.

We had apartments that needed to be rented out in May. Most renters start looking three months before their lease expiration, which meant that to fill apartments in the spring, the leasing agents needed to start selling in the heart of winter, traditionally a dead time for the rental market anywhere, but especially quiet in freezing cold Chicago.

I had to manage my expectations because, at the end of the day, until we had a real building, leasing was going to be hard. It takes a leap of imagination for the consumer who doesn't have an apartment to walk through. As a developer, you do your best by giving your leasing agents all the tools you can—including getting the word out.

In addition to a public relations firm, which put out our press releases and facilitating press stories, we hired a third-party digital marketing company, Digible. They served ads targeted to our demographic of people looking for luxury rentals in downtown Chicago. We worked with third-party brokerage companies and were paying a lot of money to be at the top of all the online rental listing services, like Zillow, Apartments.com, and Rent.com. As we leased up, we would dial back our marketing spend.

The leasing center was fully operational, staffed, and complete with collateral material, because we couldn't bring prospective renters into an active construction site for safety and liability reasons. Even if they could tour the site, without working elevators, wires hanging everywhere, and other elements of an unfinished building, it wouldn't have been practical. That's why we spent a few million dollars on the leasing center to give would-be renters as close an experience to the real thing as possible. In addition to the technology that provided information and imagery on everything from floor plans to finishes, we built out full apartments in the back. We hoped that was everything a person needed to sign a lease, barring stepping into the unit.

ROB SINGER: We officially started construction in January 2021, with completion planned for spring 2024. By summer 2023, with construction on schedule and the entire superstructure and facade nearly complete, we launched our marketing campaign with the help of Willow Bridge Property Company (our selected leasing and management firm).

The first step was to release a teaser website or landing page, which featured renderings of the completed project as well as a form that allowed prospective residents to sign up on the 1000M VIP list. This list would allow VIPs to receive information about the project and to have the first opportunity to lease units when leasing officially commenced.

By the middle of December 2023, we had over 1,300 people registered on our VIP list. We were on track to deliver the first units in May 2024. If we could sign fifty, one hundred, two hundred leases prior to opening, the building would be a dream. In January 2024 we emailed the 1,300 VIPs, inviting them to tour our new leasing center. It's common for new buildings to give VIPs a chance to learn about the product—including prices—and sign leases before the rest of the market. Only about 200 of the 1,300 made appointments to

Outdoor communal space

see the leasing center, which was disappointing. And of the 200, we only secured leases from maybe 5 or 10 people, which was even more disappointing.

It wasn't time to panic yet. The VIP page, which ended up with more than two thousand sign-ups, showed that there clearly was interest in the building. December and January are the low leasing months, and even though we had a very high-tech leasing center, no one had yet seen the inside of the building. It's hard to prelease units in the winter out of a leasing center.

Still, the building opening in May was approaching fast. We had originally projected to secure on average thirty leases every single month for twenty-four months straight, from January 2024 to December 2025, for a total of 720 units. We were already falling behind this projected average. It's one thing to write down these projections—based on data from other new building lease-ups in Chicago—but it's another thing when it's time to deliver on the projections.

In addition to the number of leases that we needed to secure, we had to meet the rental-rate projections that we had originally set forth. Failing to meet either the projected speed of leasing or the rental rate would result in a very difficult situation whereby we would not be able to find a new lender to pay off the construction loan when it was due in December 2025. There was a lot of pressure.

I maintained confidence that 1000M would be the best building in Chicago and people would recognize that once it was complete and they could see it with their own eyes. Also, we knew that demand for high-end rental was strong from the year before and that the economy was strong and that there were very few competitive new buildings.

In the short term, though, I thought that we'd have to reduce our rent prices to get momentum and/or give away free-rent incentives. Somehow, we had to get rolling.

FRANCIS GREENBURGER: We hired Willow Bridge, a full-service management company, to guide us through everything from marketing to managing residential needs. The third-largest apartment manager in the country with 180,000 units under its operation, Willow Bridge had a wide swath of areas under its purview. They not only advised us on design issues and operational budgets but were also responsible for hiring all the people that would work locally in the building.

Eventually, there would be about twenty-five full-time employees in three different divisions at 1000M (in addition to many more vendors). The management office included a general manager, an assistant manager, bookkeeper, and a resident services coordinator. To service a building of this size, where there would be many things to fix, clean, and figure out every day, a large maintenance staff would be required. In our case, Willow Bridge estimated a maintenance staff with a chief engineer and about fifteen technicians. There was also going to be two full-service concierge desks maintained by a rotation of eight concierge team members, as well as two package attendants to help with resident packages and services. It was a big team.

At this moment in the story, however, perhaps the most important department hired by Willow Bridge was the leasing team, comprised of one leasing manager and four leasing agents. These were the folks tasked with bringing 1000M to life for anyone who came into the showroom. No matter who arrived for a tour, all the agents worked with the same basic storyline: every detail within 1000M was thought out to create a curated community for you and your lifestyle. It began with a little bit of history. Envisioned by Helmut Jahn, well-known for his visionary architecture, this is his last residential building, which in itself is unique. Kara Mann's cutting-edge design approach to all the finishes and amenities spaces was unparalleled in a rental building.

The amenities package at 1000M was robust, to put it mildly. From family zones to cocktail lounges to an art gallery, the agents had a lot to work with in positioning the building as a lifestyle versus just a place to live. Part of that lifestyle is a host of services the building would eventually offer its residents when fully up and running. This went way past getting your dry cleaning delivered: 1000M would have onsite pet grooming, massage therapists, poolside snacks, short-term office rentals, and on and on.

However, from a leasing agent's perspective, perhaps the most unique aspect of the building is the fact that there are a whopping seventy-seven different floor plans from which to choose from, making the sales line "There is something for everybody" very true. A lot of thought went into the functionality of each floor plan: how the lighting is displayed throughout the unit; the built-out closet space; and the application of soft-closing integrated appliances. This is true whether in a studio or a penthouse, making 1000M's luxury approachable for a range of renters.

The building's many selling points notwithstanding, downtown Chicago had a lot of great properties and is a very competitive market. The leasing team had to market the property so that people understood why 1000M was a level above everything else they could rent in a city with a lot of options.

When I arrived in Chicago on March 1, 2024 to meet with the Willow Bridge team to hear how the leasing experience was going, twenty-nine applications had been submitted. Out of those, eight leases were fully executed. Despite assurances that the path from application to approval and signing was direct, I worried about the conversion rate of those applications. Willow Bridge, which had set its leasing goals in terms of application and not signed leases, had surpassed their target of twenty-five by March. But it still made me nervous.

Lauren Halpin, the leasing manager, reiterated that during the winter months there is a general drop in leasing traffic—and not just because of the prohibitively cold wind.

> Most assets are setting their expiration management to favor the summer months and limiting the number of explorations they'll allow because that's when most people are interested in moving. Come April through July, we'll see a very significant increase in traffic. In Chicago, most communities require a sixty-day notice, so people typically aren't looking until seventy-five to ninety days out. At that point they're working with their current community to secure a renewal or put in their notice and transfer over.

The team always knew beginning the lease-up in January was risky, and despite the tremendous effort put into hyping 1000M, the showroom was very quiet for the first week. Once the agents got past the unnerving hump of opening the doors and letting people know they were open,

Indoor communal space

they began to see signs of life. According to Lauren, they had seen a slight uptick in traffic when compared with the same period in years past. "Ten applications in the last week of February, historically one of the coldest months in Chicago," Lauren said, "It's a good sign."

Lauren and her agents were still working out how to make the tour path as efficient as possible. With a property as big as 1000M, the challenge is to give someone a full sense of the building without exhausting or overwhelming them. You need to offer a menu: a twenty-minute tour for the impatient types or one-hour tour for those who want to know every nook and cranny. "Even with the impatient ones, taking them up to seventy-three is going to be a game changer," Lauren said. "Not everyone necessarily uses the fitness center. Not everyone swims. Not everyone needs an office space to work from. But seeing that view in that incredible setting will resonate."

The eighty thousand square feet of amenities, however, were helping agents adjust the expectations of people who would be spending at the top end of their budgets to get into 1000M. Lauren explained they were able to convince some, looking for a one bedroom, to pay the same price for a studio or convertible with the argument: "If you need to get out of your apartment, there's plenty of space to do that."

The initial traffic, primarily Chicagoans already living in the downtown neighborhoods, was for the convertible studios and one bedrooms. That could have been a worrying sign for the larger units, just as we had trouble moving them when they were condos. I was looking forward to getting the perspective of 1000M's business manager, James Shortenhaus, who likened his role to an operations manager with an eye on leasing.

JAMES SHORTENHAUS

1000M Business Manager, Willow Bridge

Coincidentally, I started out in leasing in 2004 about two blocks south of 1000M at a new construction building called SKY55. For the last twelve years, I have been with Willow Bridge, most recently on two lease-ups in the Loop.

During the early lease-up phase, I am like a very intense leasing manager working with the leasing team daily. Then as we get stabilized and bring on more people, I become more of a support. Although 1000M's team will be large, I make sure it's always pointing in the right direction. We're all going to have our different personalities, but I want people to have the same messaging and branding when they talk to anyone who comes through the showroom door.

Pricing is what keeps me up at night during the lease-up. It's on all of our minds, all the time. I don't want to give units away for a lower price than I need to. We did a lot of compensation surveys to find out where our price points should sit and what people are responding to in the market. Still, we're not going to get it perfect right away. That's why we'll have daily meetings to find out what the leases are doing, so we can pivot quickly. Hopefully, we've captured the highest rent possible, but we'll raise the tiers that are moving. If we find some struggling, as far as the percentages of the building, we're going to lower those. We just recently increased a few tiers that we noticed were picking up faster than others.

Adjusting rents up or down is a constant. We look at rent growths on the renewal markets. After the lease-up phase, we switch from flat pricing to revenue-based pricing. We'll have a revenue manager, who will help us identify what the rate should be at different tiers.

Once the building opens, we will bring the amenities to life. I put amenities into two groups. There are the amenities that bring people into a building and those that keep them there. Coworking stations are a great example of a renewal amenity. Residents will stay because they have an office space they can work at five days a week. Meanwhile, the pool could look great on a tour, and you might envision yourself swimming, but it loses all its value if you never use it after you've moved in.

As the management company, we will promote and activate different areas. We create a sense of community through resident events where they can get to know those spaces and one another. We'll also find the places throughout the building that people gravitate to. I predict the lounge on the coworking floor will get heavy usage. Residents like to work in closed-off rooms, but they also like larger open spaces. With the coffee machine and the view up there in that lounge, yes, I think it will be used a lot.

FRANCIS GREENBURGER: I was anxious to learn how the spaces were used when the building opened. The first move-ins on May 1 were right around the corner. The residents would tell us if there were any problems with the programming. And if there were, I wanted to do something about it. The same was true for pricing. I wanted James and the rest of his team from Willow Bridge to feel not just free but compelled to keep the lines of communication open and running.

You all will be in the middle of the action, so you're going to know the truth," I said to the group assembled in Chicago on March 1. "The rest of us are living with concepts. My bottom-line message is: don't be afraid to be the squeaky wheel. If something isn't working, let everybody know, including me. Whether it be rents, amenities, events, vendors, we need to know how the reception is from staff and residents. We hope things will be what they are, and with transparency and straightforward communication, adjust to whatever extent we can. Don't tell us what you think we want to hear; tell us what you think we *should* know. Call me. My phone rings 24-7, so you know.

Mural by Naomi Scheck on back of 1000M

Chapter 18

"REAL ESTATE ISN'T ABOUT PROPERTY—IT'S ABOUT PEOPLE," MAY 2024

1000M Opens

FRANCIS GREENBURGER: The soothing notes of Bach's *Prelude and Fugue in C Major, BWV 846* floated softly through the eighth floor while a resident, a driver slung over his shoulder, entered the golf-simulator room and a few others in robes made their way to the sundeck to enjoy the unseasonably warm Sunday at the end of September 2024. In the piano lounge, a young woman sat at the grand piano playing the iconic progression of broken chords with the lush green of Grant Park and glittering blue of Lake Michigan as her backdrop. The piano player could have been a professional, at least to my untrained ears, but she wasn't. She was a highly skilled amateur resident enjoying one of the amenities in 1000M's eighty thousand square feet of community spaces.

With the delivery of the first apartments three months earlier, on May 1, leasing went better than predicted—much better. By May 12, two hundred leases had been signed, more than double the rate we had anticipated. As Rob said, "The demand is like nothing I could have anticipated or seen before."

LISA GRADY

1000M Tenant

I work in retail management and had opened a luxury retail store in Troy, Michigan, where I worked for a year and a half before I accepted a position at the new store the brand was opening in Chicago. Googling "luxury apartments in Chicago" and "where to live in Chicago," I began to look at buildings, hoping to find something brand-new, but nothing fit the bill. When the brand moved up my start date by a few weeks, I returned to my search with more urgency, and that's when I found 1000M. Listed as one of the top luxury apartment buildings, it was beautiful, brand-new, and in a good area.

I'm picky. I was very big on not having any carpet in my apartment, so I loved that it was all hardwood flooring. In addition, I am very particular about the bathroom layout and functionality. More specifically, I didn't want a bathtub or anything requiring a shower curtain. I was relieved the building had glass shower doors and appreciated that they didn't use the standard subway tile and were able to give the bathroom a unique design concept. One added feature was the construction of the closets in having built-in shelves and drawers, eliminating the need for additional furniture to take up space in the closet.

A feature I love about the specific apartment I chose—a twelfth-floor deluxe studio—is the kitchen island, which not only provides storage but is also simply designed, with a nice dining area that meets my needs. Moving from a larger home in Michigan to a much smaller space in Chicago, I wanted to ensure I was maximizing my space and not purchasing more than I needed; I truly appreciate the streamlined kitchen with a built-in refrigerator. If the kitchen appeared heavy with appliances, it would make the apartment feel smaller and give off less of a "home" feel.

When I moved into 1000M on May 7, it had a 2 percent residency rate. Although the building was still under construction, they did a wonderful job maintaining luxury standards by keeping the lobby clean and nondisruptive to residents. I also appreciate the "smart" concept with easy-to-use badge access to the building, amenities, and units through smartphones.

After helping me move in, my parents also fell in love with 1000M—particularly my dad, who would comment on being able to enjoy his morning coffee from the twenty-first floor. He commented that when visiting Chicago, this is the perfect place to stay thanks to the level of luxury, cleanliness, amenities, and concierge service.

Based on the overall design and concept, you truly feel as if you are at a resort, which is one word I often use to describe 1000M. From entertainment areas to the chef's kitchen, terraces overlooking the city to host guests, guest-suite accommodations, transparent vinyl igloos with furniture inside located around the pool during winter and more, 1000M does a wonderful job of making its residents feel special and at home.

FRANCIS GREENBURGER: While we all knew the fast pace of the initial leasing wouldn't keep up, it was nice to have positive feedback out of the gate. High velocity generally means the market sees a value in the product, which not only boosted our confidence but also that of our lenders and investors.

Speed and number of signed leases—not to mention satisfied residents—aren't the only barometers of success. The ultimate litmus test in the leasing-up phase is whether you are getting the rents you projected. At 1000M, the team had rented studios, a couple of penthouses, and everything in between, proving the market was also okay with our prices. Our strategy was to lease-up as quickly as possible at our original rents, even if we might be leaving 5 to 10 percent on the table. Over the next two-to-three years, we would adjust with hopefully better pricing power. Right now, however, we had a lot of operating expenses. In May, when the first apartments were delivered, there was a temporary certificate of occupancy through the forty-first floor. While interior construction would be going on until October, 1000M already had a full-time staff of twenty-two.

One of the most vital members of the team is the chief engineer in charge of keeping the building running. From the HVAC system to keyless locks to the inevitable plumbing problems, the maintenance director makes sure everything is operating. By early July, 1000M chief engineer, Arman Pasic, was mostly concerned about keeping the place pristine. With myriad luxury finishes and furniture, this was quite a job.

Arman squinted and froze in place when a young resident rode his scooter through the lobby. One scuff to the Venetian plaster finish would require the artisans who applied it by hand to return. A little nick to the curved wall of individually placed tambour-wood pieces similarly spelled disaster, according to Arman. "I wanted to put up signs warning residents," he said; but Phil Castillo put the kibosh on that. "What are we, Times Square?" the architect quipped.

As the building began to fill up—by July, 120 apartments were occupied—the eagerly anticipated final pieces of 1000M's mosaic, known as humans, were introduced into the Platonic vision of Helmut Jahn and Kara Mann. That included the pool's opening July 4; NASCAR's only street race just a few days later and only steps away; nightly fireworks from Navy Pier during the summer with a perfect view from the twentieth-floor terrace; and the upcoming Lollapalooza festival, literally in the building's front yard, otherwise known as Chicago's Grant Park.

"We're learning as we go," said Arman, while making his regular rounds through the building. Surveying the blond wood, white-upholstered chairs, and light rugs on the coworking floor, he only half joked, "Can I hose down the place in Scotchgard?"

As residents started to live in the building and put its amenities to use, Arman was watching closely to "establish routines and find our cadence." In the game room, where earlier that week he had found a pair of wet swimming trunks, an eleven-year-old resident and his friend marveled at the number of gaming controllers neatly lined up in the TV console. "Oh my God, there are so many!" the friend exclaimed.

"I know!" the young resident added. "I want to take one home."

No sooner had the words come out of the child's mouth than Arman texted a maintenance tech. Within three minutes, the tech stood in the game room where Arman pointed at the controllers and said, "We need to move these behind the front immediately and figure out some kind of check-out system."

ARMAN PASIC

Chief Engineer, 1000M

I grew up in the bowels of a power plant in Bosnia with my dad who was a mechanical engineer. I was fascinated by this huge structure at the edge of town, with its smoke stacks and cooling towers. When I was around ten, I asked my dad to take me to work, and he did. I was curious about structures, from inception to construction—how they operate and who takes care of them. Though they may be inanimate objects, to me they are living and breathing, each with a unique personality.

I've been in this industry for almost two decades. I helped open and operated two residential towers on Clark Street, three blocks away from 1000M. When McHugh started the foundation work on the skyscraper, I called my engineering manager and said, "Hey, if we ever get the contract on that building, I'd like to throw my name in the hat." Even though it wasn't a straight line from Clark Street to here, I'm here now and grateful for the opportunity. From Helmut Jahn to Francis Greenburger to this iconic address on Michigan Avenue, 1000M is a landmark in and of itself already. To be attached

Arman Pasic

to the project means a lot to me; I never give up an opportunity to mention it to my kids.

When it comes to preventative maintenance, we started with the basics: what is our HVAC system design? What about the plumbing system? We have a tower of seventy-four floors. How many different mechanical and plumbing zones are there? Where are the booster pumps, isolation valves? Working closely with the subcontractors and McHugh, we learned about the system's design. I walk around with my little notebook and make sketches. I'm not shy about asking questions, meeting with pipe fitters and plumbers and foremen daily. I am also lucky to have the legendary engineer Tom Gilbertson guide me through the mechanical and plumbing systems. Even today, Tom calls often to check in and is generous with his knowledge. I cherish those calls.

The HVAC system's heating boilers, pumps, and cooling towers are systems we've encountered before. With the plumbing, we know how it is supposed to operate in theory. But how it behaves once we have users in the building is extremely important. Does every shower have enough pressure? Is the water temperature right? I worry about everything that most people take for granted. Will this shower door hold the water back? Are you going to step out into a puddle on a floor with no drain and slip? Those are the things that I worry about. I get granular. How many labor hours does it take to pull the recycling out? What's the most efficient path? When should we close the pool for cleaning?

When it comes to the interior finishes, the first thing I see, even in a place as beautiful as 1000M, is all the work it may take to keep everything as it is supposed to be: clean and fresh as the architects and designers intended. The art, the skill, and the craft are lost on me in that initial approach, because I'm thinking, "What's the finish on the elevator? Does it take fingerprints?" It takes a little while for me to digest it, so that I too can then appreciate the aesthetics of the space.

The quality of the preventative maintenance will only be as good as the people we have been doing the work, whether that's my assistant chief engineer, me, or the janitorial staff (the first rung of people to touch everything the residents use every day). As the building fills up, it will be incumbent on me and my colleagues to be the best service team so that, come renewal time, there will be no reason for them to leave. That means anticipating issues before the resident even knows it's an issue. I lifted a line from the mission statement of the Ritz-Carlton Hotel Company: "We anticipate and fulfill even the unexpressed wishes and needs of our guests." It takes dedication and a service attitude. The enormity of the task and responsibility of being the chief engineer can keep me up at night, but it truly is a calling.

FRANCIS GREENBURGER: Arman and his team weren't only learning how the HVAC, plumbing, and 1000M's other mechanical systems behave once "users" were in the building. LIVLY, a resident app and property management platform, allow people to put in maintenance tickets and communicate with management about any issues. This provides valuable insight about how the spaces are being utilized, where they need to allocate more personnel, and other issues. "Somebody put in a work order to do their dishes after they hosted a party in one of the lounges," said Arman, who explained that was not part of the amenities package but proceeded to help the resident clean up anyway. Some feedback from residents, however, proved helpful in making the place better. One resident sent a five-paragraph work order for the gym in which she scrutinized every piece of equipment in the two-floor gym, clearly demonstrating mastery of the subject. Arman immediately called the vendor's technician out to the building: "Whoever this woman is, we should thank her. She knows her stuff, and we have a standard to uphold."

There are always the unforeseen hiccups, like the keyless doors. Some residents—most of whom used an app on their smartphone that functioned as a key—couldn't open their doors, while others couldn't lock them. The engineers worked through that, but another problem that couldn't be fixed with technical expertise was what happens if someone loses their phone—say during a night out. The staff doesn't have a way to get into every apartment. If you return home at 2 a.m. after a night at the bars, you will be waiting in the lobby until an engineer arrives in the morning. The overnight concierge already had experienced that scenario multiple times.

Overall, however, it was amazing to see the space in use. Sometimes buildings put in all these amenities or communal spaces, and people don't engage. If anything, 1000M was seeing *too* much engagement. Residents were so excited about the Fourth of July, when the pools opened, that many invited upward of ten guests. At single-digit occupancy rates, the staff managed. But they quickly needed to implement a guest limit per unit for the amenities.

From the management offices on the second floor, James and Armen met weekly to go over the feedback from the leasing, amenities, and maintenance staff to troubleshoot any issues. Whether it was the chief engineer, the janitor wiping down the windows, or the resident-experience coordinator, I saw employees with experience and noble intentions, doing their jobs to the best of their abilities. It is wonderful to have people who carry my same pride in this building. From my point of view, the road was certainly long to get here, but it doesn't end with the construction. If bringing 1000M to life is not realized properly, then whatever may have preceded is wasted.

One of the most front-facing staff members that allows 1000M to operate on such a high level is the chief concierge. Scott Shiff—a veteran residential concierge for the last twenty-two years in San Francisco, Las Vegas, St. Louis, and finally in Chicago—opened the building in late April.

SCOTT SHIFF

Chief Concierge, 1000M

I got into the guest-services profession quite accidentally after many years as a professional sports reporter. My first job was at the Four Seasons Residences in San Francisco. After five interviews and despite having no previous experience in luxury hospitality, they hired me because they sensed I had the personality to service VIP clientele with the utmost care. In my previous career as a radio and print sports reporter, I needed a high-level of attention to detail and the ability to communicate effectively. Those same skills are crucial to being a concierge. My job is predicated on making things run seamlessly, ensuring that they are done correctly, the first time, and in a timely manner. I am in charge of a staff of up to ten other concierges, and my duties include training them to deliver the same level of service as me.

Certainly, by height and number of units, 1000M is nearly double the size of any building in which I've previously worked, and from an amenity perspective, this is by far the most luxurious. For example, a

Scott Shiff

building in St. Louis I worked at had all of their amenities—the pool, bar, pool table, giant TV—on one floor. 1000M has seven different floors of amenity space.

A typical day for a concierge remains the same from building to building. Of course we greet residents, their friends, and vendors coming in and out. Considering the sheer number of people that come to the property on a daily basis, I take on the roles of both chief concierge and head of security. We have over ninety security cameras that we constantly monitor. If we see anything out of the ordinary, we react and take care of it immediately. However, my professional motto is to be prepared and anticipate what might happen, so we can prevent issues before they become problems that impact our residents. We must always be alert and address any person that we don't recognize and not let them up to the floors without permission. Because their keys are digital on their phone, residents can give a key to anyone they wish. We want to make all visitors feel welcome here, while protecting the building's integrity and our residents' safety.

One of the biggest and most time-consuming aspects to running a luxury residence is facilitating the delivery of packages. Amazon, UPS, and FedEx literally come all day long. These guys show up from the moment I arrive and continue to deliver regularly all through the day and night. Although we have a separate package department, when they close in the evening, our staff is here to help residents retrieve their deliveries. The building handles several hundred packages a day.

Ever since I began my career in residential hospitality, I have enjoyed using my background in creative writing to produce and distribute a weekly newsletter, with my recommendations about events going on around town. I also facilitate our Pet of the Month contest. Additionally, I host a monthly trivia night for the residents to come have fun and get to know each other.

I have been trained to never say no to anybody, even if I know the answer is no. Instead, I ask questions and problem solve to find a solution that works for everyone. Residents are sharing this space with more than one thousand other people, and the concierge staff will enforce rules to ensure the enjoyment of the property by all who live here and their guests. For example, a resident kept parking in a space designated for visitors. I asked her why she would do that when she had her own space. It turned out she had an electric vehicle and needed a space closer to a charging station. We found another one that fit her needs.

For two years, I was a butler for VIP high rollers in Las Vegas. My boss, who had been a master butler for over twenty years, servicing some of the most famous and high-profile clients you can imagine, taught me that this role is like being an actor on a stage. Whatever your problems are, you must leave them at home. There cannot be any negativity or impatience detected by our residents. We are here to provide exceptional customer service: think on our feet, be confident in our decisions, and be clear in our communication. My job is to help people, and I love doing so. My reward is the feedback I get from my clients, who appreciate my efforts.

FRANCIS GREENBURGER: We delivered a tremendous building with popular amenities—on time and on budget—but in real estate there is always another hurdle. By early November, I was alarmed after Rob said we were running tight at current interest rates with respect to having enough remaining reserves to cover interest payments in the spring.

The budget is set up to carry the building for a certain period. In an ideal world, by the time reserves run out, you have leased enough to cover your expenses.

In Casa Mara, leasing moved ahead of schedule, so we ended up with a surplus. Leasing in another development in Panama City took longer, so we had to pay a portion of the interest with equity that we

Bird's-eye view of 1000M

didn't expect. In Chicago, the intersection of income and expenses was still evolving.

We had delivered 1000M into a robust rental market. The Midwest—which the commercial real estate website Globest reported "challenges Miami for hottest rental market"—was becoming a desirable place to live in part because of the region's economic mix of tech and manufacturing. Chicago had seen meteoric rent growth, with a 32 percent year-over-year high in 2021. By fall 2024, downtown Chicago's rent growth had slowed, particularly for high-end buildings like 1000M, but demand was still strong. "The stagnation is largely due to the influx of new supply, with 3,582 units expected to be delivered by the end the year, marking the largest annual addition since 2017," *The Real Deal* reported at the end of September. "With only five hundred new units expected downtown next year, supply is set to dwindle, potentially pushing rents up again." Experts were predicting rent increases of up to 5 percent.

An outdoor terrace on level 20

But if I had a quarter for every time an expert was wrong, I could erect many, many more buildings.

By the end of fall, 1000M was 60 percent rented. Our payments of $1 million a month on the construction loan meant we were running a deficit of about $500,000 a month. By March, we needed to have enough income from rents to cover the expenses. Unfortunately, the rental season in Chicago didn't start up again in any meaningful way until March. Thanks to many moving pieces, we didn't have all the facts. Still, Rob was intuitively concerned. This had the potential of becoming a $5 million problem.

Before we could pull off the last trick, replace the construction loan with a mortgage, the building had to be leased up. Certain banks will give you a new mortgage if the building is 80 percent leased, taking it on faith that you will get up to 90 percent. Other lenders want you to have 92 to 93 percent occupancy. A mortgage wouldn't solve all our problems. We still needed enough leasing income to pay the monthly mortgage and building expenses, from management fees to the utilities. That didn't mean the fixed mortgage couldn't *cause* problems. One percentage rate difference on a $300 million mortgage is $3 million annually, and if it's a fixed rate for ten years, the percentage point is quite significant.

The goal wasn't just to keep the lights on. Depending on the mortgage we eventually received, the hope was to return extra money to our investors. While we needed $305 million to cover the development costs, our goal was to get a mortgage of $350 million. This wouldn't be paid to large banks or corporations but rather individuals like Steve Poliseno, an Italian immigrant who sunk every penny from the property sale of his beloved Astoria Sports Complex into 1000M as part of the 1031 financing technique I invented for the building.

The tax-deferred transactions that allow real estate investors to defer capital gains taxes by reinvesting the proceeds into a replacement property, which I figured out could also be employed in a major development, had been so successful in Chicago it gave me confidence to enter into another mega deal. At the end of March, I shook hands on a mixed-used project with more than nine hundred apartments and 22,000 square feet of retail space in the heart of Boynton Beach, a hot area of Florida located between West Palm Beach and Delray.

When we closed on the land in Boynton, Deutsche Bank gave us a great loan that also came directly from the experience of working on 1000M. Originally, we had an inferior loan from another bank for the Florida project. However, two Deutsche Bank bankers came out to Chicago to check on the progress of the building and were so pleased they expressed a desire to do more business together, so I gave them the opportunity to give us a better deal on Boynton—and they did.

It was great to have the confidence of one of the largest banks in the world, but it was moving that someone like Steve Poliseno would put so much at stake personally by investing in 1000M. For him, this wasn't an abstract financial transaction; he had entrusted his children's and grandchildren's inheritance to us.

STEVE POLISENO

Investor, 1000M

This was the first investment I made in my whole life. Well, actually, the second. The first investment was purchasing an old, abandoned icehouse in Astoria, Queens, in 1976.

Until the 1930s, refrigeration was not widespread. These ice plants were like Con Edison today, supplying ice to all the houses, restaurants, shipyards, airports, you name it. Companies, like the one that owned the icehouse in Queens, never imagined ice wouldn't be a necessity. My brother was working there when the business went bankrupt. The ice business wasn't easy. When my brother Nicolo, a father of five, was given the devastating news, Commandatore Mercurio, one of his supervisors, took him aside. "Nick, you don't have a job anymore," said the man, who was part of our Italian community. "Listen, I know your father, Giuseppe, has a little money and your younger brother, Steve, has a good head on his shoulders. Talk to them about buying this place."

We bought the building at auction from the city and turned it into a sports complex. I grew the Astoria Sports Complex, dubbed the Madison Square Garden of Queens, from a twenty-five thousand square foot dilapidated icehouse into a seventy thousand square foot facility with soccer fields, batting cages, a swimming pool, a gym, and a birthday party room. We had everything in there! There was a catering facility and indoor parking. The business was tough—we almost went bankrupt a few times—but to this day, I don't have Christmas Day dinner until I have personally placed a wreath on the grave of the supervisor, Comm. Mercurio, who told my brother we should buy the building. He changed the future of my family. Sadly, I lost my brother in 1996 and then my father four years later, but since then I have run the complex with my children, Giuseppe Jr. Apollonia and Vittorio.

Steve Poliseno with Francis Greenburger

When the pandemic hit, I had to get out. While we had saved the business on many occasions, this time it was going to financially ruin me. Fortunately, we were able to sell the building. But it wasn't without regret and sadness. My biggest worry had to do with my three adult children, who all worked at the complex. I was in my seventies, but what about my poor kids? They were newly married and starting families. Any decision I made was not just for me but also for them.

Hoping to find a 1031 exchange, we reached out to Marilee Hill, a financial advisor at Emerson Equity, who put us in touch with TEI, which had a development project in Chicago. In October 2021, we sold the building and had 180 days to reinvest the money. We met with TEI's head of equity, David Becker, who was very nice and convincing, but I'm the old-school type: I want to see what I'm going to get into.

David, my son, Vittorio, and I traveled to Chicago. There was a model of 1000M, but the site was just dirt. I thought, "This is very far-fetched." I've always been leery about scams and people trying to take advantage. As I said, this wasn't only about me. My daughter and daughter-in-law were both pregnant. I had to make sure I was able to protect my whole family. David worked his butt off to convince me that these guys were trustworthy, but this was all new to me. The amount from the sale was more money than I could ever imagine possessing. Nervous about putting all my eggs in one basket, I would have felt better if I could have spread the money around, investing $5 million into various properties. That wasn't possible, however. I had no choice but to invest the whole amount.

That's where Mr. Greenburger comes in. He walked into the room when we were in his Manhattan offices, and David said, "This is Francis Greenburger. This is the owner of TEI." When we shook hands, I felt his honor, sincerity, and strength. Something said to me, "This is the Real McCoy." Mr. Greenburger convinced me that if in three years' time I wanted to move some of the money out of 1000M, I could do that.

Ribbon Cutting, June 13, 2024

I had financial advisors and attorneys telling me, "You're out of your mind putting every penny you own into 1000M."

However, Mr. Greenburger proved them wrong.

We started getting distributions from TEI the very first quarter. Nevertheless, I kept a constant eye on 1000M. I couldn't trust anybody. When David Becker, my son Vittorio, and I visited the site in Chicago in 2021 while I was considering the deal, my son and I went to a steak restaurant on the corner across the street from the building and made friends with a couple of waitresses and the manager. Periodically, I would ask for them to send pictures of the construction.

Riding in an Uber to the opening ceremony of 1000M in late June 2024, I couldn't see any skyscrapers from the backseat window. All of a sudden, we pulled up and I got out of the car. Looking up, I saw this monster of a building that just kept on going up forever. The more I looked up, the higher the building went, as if it could touch Heaven. "Wait a minute," I thought to myself. "The driver used the address 1000 South Michigan Avenue in his maps." I had the wrong building!

It took a little convincing from the driver, who explained that the building's name is 1000M, but its address is 1000 South Michigan Avenue. Words are difficult to describe what I felt in that moment. It was a lot of emotion—a sense of accomplishment, and gratitude to God. "Wow!" I thought. "If only I could show this to my father and brother. It reminded me of when my children were born.

I was born in Italy and came to this country when I was nine years old. It was hard to leave my friends, family, and everything I had ever known. At the age of fourteen, I was diagnosed with a brain tumor and was told I would not reach my twenty-fifth birthday. I'll never forget when my father, who arrived here in 1956 with nothing in his pocket, said, "We are going to make it."

Sure enough, here I am.

FRANCIS GREENBURGER: From individual nest eggs to major urban skylines, there is so much riding on the success or failure of a development project. People often ask me how I sleep at night. Sometimes well and sometimes not as well. I know there are always going to be problems, and the higher the building, the bigger they grow. But I operate from a place of confidence thanks to an extraordinary group of colleagues.

It begins with the team at TEI, many of whom I have worked with since the 1980s, and a whole new generation who will continue our legacy. From mortgage analysts to asset managers, everyone brings expertise, experience, and commitment to their job. The same is true for the team that put 1000M together. The architects, interior designers, construction management and crew, marketing professionals, engineers, accountants, brokers, electricians, plumbers, gardeners, urban planners, lawyers, janitors—the list goes on and on—everybody has performed. I don't think anyone wishes the design, unit mix, amenities, or staff were different, and that speaks to hiring the best team. Naturally, everyone thinks they have the best team when they hire them. You don't really

PARK HOTEL
EBONY

find out if the team is truly the best until you're done. As we sit here today, I'm pretty sure we did hire the best team.

While the buck stops with me, the developer, I'm not alone. With these colleagues, I believe most problems can be solved. I say "most" because there are a lot of variables in determining the outcome that we can't control. When we were building 50 West, the Helmut Jahn tower in lower Manhattan, the glass for the windows arrived cloudy and had to be sent back to China, helping to run up our budget by many millions. A global pandemic stalled construction on 1000M. No amount of worrying can help predict the many unforeseen scenarios that arise in all our lives. With that ethos, my definition of success is doing the best you can within the circumstances. That's how I guide myself, including doing the very best I can to surround myself with people who have similar intentions. Time and again, I have found that residents respect and respond to that kind of effort.

In one of the early iterations of 1000M, there were different amenities areas depending on whether you lived on the higher, more expensive floors, or lower down. It has been interesting to watch the evolution of the culture of 50 West, which also has a robust amenities program. It is a hyper-social building in a way that defies the norm in New York City. In Manhattan, it's typical to live in a building for thirty years and not know anybody. In 50 West, I'm constantly surprised that everybody knows everybody, and they want to spend time together. When the building was going to have a party on July Fourth, I told the management that it was never going to work: "These people all have beach houses, country houses. Nobody's going to come to a July Fourth event." Boy was I wrong. A hundred people showed up in a building of only 191 units.

In a typical residential building, you live in your apartment and that's it. Where do people meet, except in the elevator? The amenities and common spaces in buildings foster social integration, or put in a simpler way, life. That's why I decided, much to the chagrin of some of our consultants, that the top of 1000M had to be open to everybody.

Piano lounge

The seventy-third floor's observation deck—higher than any other in a residence in the city—is essential to the building. Residents can share with friends and one another the otherworldly experience of the Chicago elements, snow or a warm summer breeze, blowing through open windows where the view of the city goes on forever.

1000M is, in essence, a shared experience. You have your apartment, and then you have the whole world. You can join a spin class in the gym, take a meeting on level 21's coworking floor, or sip mimosas with neighbors at Sunday brunch in the piano lounge.

Elise Zhang was taking advantage of the lack of a Sunday brunch on that late September day to practice piano. She rented an apartment in 1000M back in June because it was the only building with a grand piano as an amenity. "I have a digital piano upstairs," she explained. "But a real piano, especially a grand piano, feels different."

A Guangzhou native who came to Chicago for college and currently works in marketing for a tech company, Elise started piano when she was three years old. "I stopped playing piano around thirteen years old, because school in China is very hard and takes all your time," she said.

"Now that I have more free time, I have picked it back up as a way to calm myself down."

At first practicing in public was understandably uncomfortable. While Elise is an accomplished pianist, she wasn't comfortable having others hear her make mistakes (even if she was the only one who knew she had made a mistake). People interrupting her to ask questions or make comments was also jarring. "It used to make me nervous," she said. But by continuing to practice in the communal lounge, what started as unsettling gradually became inspiring. "When people are on their way to play sim golf, sometimes they stop to listen. It's nice," she said, adding, "I find inner peace while playing here."

The prelude Elise played built slowly and methodically. In deep concentration, she moved her hands lightly across the piano keys, unbothered by the nearby presence of a large brightly colored piece of art. The artist Judy Pfaff gave the human-sized piece, a nest of honeycomb cardboard, expanded foam, plastics, and fluorescent light, the title *Time is Another River*, inspired by a poem by Jorge Luis Borges.

"Remember Time is another river," Borges writes. "To know we stray like a river/and our faces vanish like water."

In the poem titled "The Art of Poetry," the Argentinian writer describes what it is to "convert the outrage of the years into a music, a sound, and a symbol."

"Art is endless like a river flowing," begins the poem's last stanza, "passing, yet remaining."

Like a painting, where the artist combines color and brushstroke to create an image, or music, a careful arrangement of notes and rhythms played by musicians, a skyscraper is also a composition of meticulously choreographed design and materials, brought to life by a talented orchestra of designers, construction workers, and construction managers.

ACKNOWLEDGEMENTS

First, special thanks and recognition to R.S. whose name I am not allowed to mention for some reason.

To my wife and life partner, Isabelle, thank you for all your wisdom, design input, and support in regard to 1000M and all my—and our—many other crazy dreams.

To my writing partner and friend, Rebecca Paley: where will our next adventure take us… my bag is packed.

To my partners, Jerry and Jordan Karlik, Mindy Ouyang and Eli Abubekar, and Bob Kantor: thank you for your openness to seeking "beauty, best in class, and 'glory'" with me.

To Helmut Jahn, Phil Castillo, and Lynda Dossey: thank you for always being willing to "make it better."

To Kara Mann and team for your patience with my "suggestions."

To Patty McHugh and her team for outstanding leadership and professionalism.

To my bankers Rod Colburn, Stephen Reinhard, and Anthony Valvo: thank you sharing in the vision and believing in us.

To David Becker for taking a chance and convincing others to believe.

To my publisher Colin Robinson for his enthusiasm, dedication, and flexibility in publishing this book.

To Phil Brody (in memoriam) and all my many colleagues at TEI who made this happen each step of the way.

To my son Noah who already knows more about construction than I do, and to Claire, Julia and Morgan for being enthusiastic in their support of 1000M.

To my readers who have invested quite a bit of time to reach this point in the book, feel free to contact me at fgreenburger@timeequities.com with any feedback or questions.